101 PROBLEMS for the ARMCHAIR SCIENTIST

First edition published in 2017 by Quid Publishing

Conceived, designed and produced by
Quid Publishing, an imprint of The Quarto Group.
The Old Brewery, 6 Blundell Street,
London N7 9BH, United Kingdom
T (0) 20 7700 6700 F (0)20 7700 8066
www.QuartoKnows.com

ISBN: 978-0-85762-499-4

Layout by Tony Seddon
Illustrations and cover by Matthew Windsor

Printed in China

FSC
www.fsc.org
MIX
Paper from responsible sources
FSC® C017606

101 PROBLEMS for the ARMCHAIR SCIENTIST

Such as

HOW DO HEADPHONE WIRES GET SO TANGLED?

JOEL LEVY

CONTENTS

INTRODUCTION

'Mother-Wit comes from all kinds
of experiences,
Of birds and beasts and of tests both
true and false.'

– William Langland, *Piers Plowman* (trans. Terence Tiller, 1981)

Science is seen today, and has been seen throughout its history, as remote, esoteric, obscure; the pursuit of elites and scholars, protected by jargon, institutions, complex mathematics and even more complex technology. Today science seems fantastically specialised and astronomically expensive, with colossal colliders searching for fundamental particles, and international research efforts decoding the infinite complexity of the genome and the brain. Even at its inception, science was the province of gentlemen scholars, and its practitioners wrote in Latin and in the language of mathematics. Dismissing criticism that his landmark *Principia* was impenetrable, Sir Isaac Newton (1642–1727) retorted, 'To avoid being baited by little smatterers in mathematics, I designedly made the *Principia* abstract.'

This book is for the smatterers, welcomes baiting and has designedly been made accessible. Its intention is to present common questions, everyday problems and fascinating issues in a way that makes them approachable, comprehensible and even fun. Science is not confined within ivory towers and remote institutions; it is in, on and around you every second of the day, from the commonplace 'I wonder why?' to the teasing 'What if?' Indeed, one of the most appealing aspects of science is that it can be accessible, open and democratic; and, notwithstanding the intellectual snobbery displayed by Newton, this was the animating spirit of its birth. The motto of the Royal

Society, crucible of the Scientific Revolution, is effectively, 'Take no one's word for it'. The essence of science is the drive to test for oneself the truth of propositions. The history of science is full of engagingly simple and straightforward experiments that reveal profound truths and laws – and, wonderfully, many of them can be done without even leaving your armchair.

How to use this book

The 101 problems and questions in this book are divided into five themed sections, ranging from the science of the domestic, personal and everyday sphere to outer space; from the natural world and the nature of being human to the fundamental principles of the universe. Each entry is illustrated with a short scenario, together with investigatory notes, interesting asides and concise explanations. The scenarios run the gamut from the prosaic to the fantastical, exploring ways of presenting complex ideas and underlying principles, so you can look at them with fresh eyes, grasp accessible analogies, test counterfactuals and reframe questions in more productive ways. The treatments are short and do not attempt to be comprehensive, although where possible they try to be definitive; the aim is to provoke argument and guide thought – to provide a kind of sandpit for testing and investigating scientific ideas, which can be explored without getting out of your seat. The mind transforms the humble armchair into a laboratory for thought experiments, and your imagination is the only apparatus you will need.

'Imagination is more important than knowledge. Knowledge is limited. Imagination encircles the world.'

– Albert Einstein (1879–1955)

EVERYDAY LIFE

'"Science" . . . is routinely practiced not only by physicists, chemists, and biologists, but also by historians, detectives, plumbers, and indeed all human beings in (some aspects of) our daily lives.'

– Alan Sokal, *Beyond the Hoax: Science, Philosophy, and Culture* (2008)

001 GET KNOTTED

As his morning train pulls into the station, Matt pauses his music, carefully coils the cord of his earbuds and delicately places them in his pocket. This time, he thinks, he's going to make sure they stay out of trouble. But when the end of the day rolls around and he pulls them out again, ready for his journey home, he finds them hopelessly tangled like the legendary Gordian knot that Alexander the Great could only slice through with his sword.

WHY DO EARBUD CORDS GET TANGLED? WHAT IS THE PHYSICS BEHIND SPONTANEOUS TANGLING AND KNOTTING OF CORDS, AND IS THERE ANYTHING YOU CAN YOU DO TO PREVENT IT?

Navel knots

One cord that mustn't get knotted is the umbilical cord, since knots can choke off blood and oxygen supply to the baby *in utero*. Fortunately, this is a very rare phenomenon, happening only in about 1 percent of births. Umbilical cords are relatively thick and are confined within a very tight space, which helps to reduce the probability of spontaneous knotting.

Knot likely

This problem has fascinated mathematicians for hundreds of years. The branch of mathematics known as knot theory has been of special interest to scientists working with strings or chains on the microscopic level, such as molecular biologists researching protein chains. Take a string or chain and shake it about: Given enough time, and assuming that the length of the string is above a critical threshold, a knot will spontaneously form. Once a knot has formed, it cannot disappear unless it is shaken off the end of the string – the knot is said to have 'topological stability'. So as soon as there is even a small probability of a knot forming, the longer the string is shaken, the more likely it is that a knot will appear.

Spontaneous knotting

A 2007 paper, 'Spontaneous Knotting of an Agitated String', by Dorian Raymer and Douglas Smith, discussed their findings after putting coils of string of various lengths into a box and shaking it. They found that, for strings longer than a critical length of 46cm (18 inches), the probability of knot formation started to increase, reaching a plateau at around 150cm (60 inches). Earbuds for a smartphone typically have cords 60 inches long. Raymer and Smith also found that stiffer strings and a smaller box reduced the probability of tangling, suggesting that one simple tactic to reduce tangling is to keep the cord in a small pocket or bag.

'How to tie the strongest knot ever:
1) Put some headphones in your pocket
2) Wait one minute.'

– Bill Murray (b. 1950)

002 CEREAL ATTRACTION

Steve is trying to get the kids ready for school. They won't stop running around, and it takes half an hour to get them to sit at the breakfast table. 'Look,' he says to encourage them, 'I got new cereal. It's Sugar-Frosted Crunchy Hoops!' The children clamour for a bowl of the frosted confection. Steve pours milk into their bowls and gives them each a spoon, and then goes to get dressed.

Fifteen minutes later he calls for them to come and get their shoes on, but there is no response. The kids are still at the table, staring in fascination at their cereal bowls. 'Come on, kids,' says Steve, but he is greeted with silence. 'What are you looking at?' He peers over their shoulders. His eldest has eaten all the cereal except for two hoops, and she keeps picking the two pieces out of the bowl, dropping them back in a little distance apart from one another, and watching as they glide towards each other and cosy up together. 'You see, Dad,' she says, 'they keep sticking together. Are they magnetic?'

WHY DO THE PIECES OF CEREAL IN A BOWL OF MILK STICK TOGETHER?

Floaters

The peculiar behaviour of floating cereal in milk is down to a combination of buoyancy and surface tension. The Sugar-Frosted Crunchy Hoops, like other similar cereals, are made from cereal paste extruded at high pressure to give shapes of light, foamy dough, which are baked into light, foamy cereal pieces. Since they have very low density they are extremely buoyant and float at the top of the milk, where they fall prey to the effects of surface tension.

Surface tension

Water (and milk is mostly water) has many unusual properties for a molecule of its size, and many of them are linked to the way that electrons arrange themselves across the molecule, giving it polarity (opposing charges at each end). This means that each molecule can act like a tiny magnet, and the attractions between the hydrogen atoms on one water molecule and the oxygen ones on another are called hydrogen bonds. Each water molecule feels the pull of other water molecules in all directions around it, except for those at the surface of a body of water. These only get pulled from the side or below, because there are no water molecules above them, and this makes the surface of a body of water slightly resistant to deformation, almost as though it had a skin: this is surface tension.

Down in the valley

Surface tension makes it possible for lightweight insects to skate on the surface of a pond, and it means that floating cereal pieces sit in a very shallow bowl or depression on the surface, a bit like a ball sitting on a rubber sheet. Each Sugar-Frosted Crunchy Hoop thus sits in its own tiny valley, and when two of these valleys come near to each other they tend to make the hoops slide down to the shared centre. Once at the bottom of the valley, the hoops cannot get out and so they tend to clump together. Other passing hoops will similarly be drawn into the valley.

003 BLUNDERBOT

When Roger was a kid, he read one sci-fi story after another in which humans had robot companions or servants. He couldn't wait to grow up and get his own robot butler. But now Roger is an adult and robot butlers are nowhere to be seen. The best that technology can offer is an overgrown hockey puck that very slowly cleans your carpet or wanders aimlessly around your garden. Fed up with waiting for someone else to invent the robot butler, he decides to do it himself.

After six months of hard toil he is ready to show his robot butler to the world. Technology journalists from all over the world gather to see the demonstration. Unfortunately, it doesn't go well. The RoboButler bumps into things; fails to understand spoken commands; trips over anything in its way; drops things it is carrying; gets lost; can't get upstairs; falls over, breaks down and needs to be fixed; and finally, after just 20 minutes of ineptitude, runs out of power and switches off. Roger has to admit that the future is a long way off.

EVER SINCE THE GOLDEN AGE OF SCIENCE FICTION IN THE 1940s AND '50s, ROBOT SERVANTS HAVE BEEN A MAINSTAY OF VISIONS OF THE FUTURE. SO WHY DO WE STILL NOT HAVE ROBOT BUTLERS?

Robots aren't working

As Roger's prototype has demonstrated, many of the tasks that a robot butler would be required to fulfill are currently too difficult for robots to master. A successful robot butler would have to be agile, robust, dexterous, resilient, resourceful and reliable. Above all, it would need to have enough power to keep going without being plugged into the mains. Today, some robots can do some of these things, but getting a robot to combine these attributes, and in particular to operate with a fair degree of autonomy, is beyond the reach of technology. Some robots are able to navigate complex spaces and respond to changing conditions in real time. Some can walk up and down stairs and get up when they fall over. Some can pick things up without breaking them. But few of these are robust enough to be allowed out of a laboratory, or cheap enough to be genuine mass-market items. Most problematically, battery technology is not sufficiently advanced to power such a device for more than a few hours at most.

RoboBeast

Perhaps the closest invention to the robot servant of fiction has been the robot created by Boston Dynamics for the US Marines: the Legged Squad Support System (LS3), a field version of a robot called Big Dog. Intended to serve as a mechanised beast of burden, the LS3 is able to carry heavy loads, obey simple instructions (such as 'follow' or 'stay'), cope with rough terrain and get up if knocked down. But it was rejected for active service because of major drawbacks: To power such a device it was necessary to have noisy motors that would attract the enemy's attention, and it was simply not economical to put such an expensive asset into the field.

004 TOUGH CRITTERS

Rupert has heard the popular legend that cockroaches will rule the Earth after nuclear Armageddon has wiped out humanity, thanks to their radiation-resistant hardiness. He has also heard a related bit of folk wisdom, which holds that you cannot microwave an ant. Having nothing better to do one day, Rupert decides to try it out for himself. Scooping up a hapless and unsuspecting ant from the backyard, he drops it squarely in the centre of the microwave, closes the door and switches it on for a minute.

When Rupert presses that switch it causes a magnetron – a microwave generator – on the right side of the oven to send a stream of radio waves with a relatively short wavelength (hence 'micro') into the interior. These electromagnetic waves pass unhindered through air and materials such as plastic, paper and glass, but they are heavily absorbed by molecules of water and fat. When a water molecule absorbs the energetic microwaves it begins to

HOW IS IT THAT A KERNEL OF CORN WILL POP AFTER JUST A FEW SECONDS OF MICROWAVING, YET AN ANT, OF ROUGHLY EQUIVALENT SIZE, CAN SURVIVE APPARENTLY UNHARMED INSIDE A MICROWAVE OVEN?

vibrate at a rate of nearly 5 billion times a second. Friction between rapidly vibrating water molecules generates intense heat, as the inventor of the microwave, Percy Spencer (1894–1970), discovered when experimenting with a magnetron during World War II. He noticed that a chocolate bar in his pocket melted when exposed to microwaves, and followed up this observation by placing first a bag of corn kernels in the path of the microwaves (thus becoming the first person to make microwave popcorn), and then an egg, which exploded just as a curious engineer bent down to look at it. Surely such a fate awaits Rupert's unfortunate experimental subject? In fact, when the timer pings and he opens the door, he sees that the ant is unharmed, just as he had hoped.

Standing waves

One reason the ant is safe and sound is that it was free to roam about. Inside the oven, the microwaves blast out from the magnetron on the right and bounce back from the opposite wall. In some places the peaks of the waves coincide to create standing waves of high-intensity heating, but in others the troughs cancel out the peaks to create zones where there is no heating. In an average microwave oven these hot and cold spots alternate roughly every 7.5cm (3 inches), and this is why most microwave ovens have revolving plates, to ensure that food is evenly heated. An ant is small enough to sit comfortably within a 'cool zone' and can simply move away if it finds itself in a hot zone.

Cool customer

Another reason that ants seem immune to microwaving is that their small size makes it easy for them to cool down. Small creatures have a higher ratio of surface area to volume, meaning that they lose heat more quickly. In a microwave oven the air itself is not heated, so if the ant does pick up heat from the microwaves it can quickly radiate it to the surrounding air.

005 HOT AND COLD

The year is 1963 and Erasto B. Mpemba, a 13-year-old schoolboy in Tanzania, and a classmate (we'll call him Abasi) are making ice cream in their home economics class. Erasto heats up some milk in order to dissolve sugar into it. The recipe states that the mixture should be left to cool before putting it in the freezer. However, space in the cool box is limited and Abasi gets in ahead of Erasto by not bothering to heat his milk; instead, he puts his cold mix straight into the freezer. Not wanting to be left behind, Erasto decides not to wait for his mix to cool and places his still-steaming mixture straight into the freezer.

An hour and a half later, Erasto is surprised to discover that his ice cream has frozen solid while Abasi's is still gooey. Erasto relates this curious observation to his teacher, who insists that he must be mistaken: Obviously a warm liquid cannot freeze before a cold one. Undaunted, Erasto repeats the process, with the same result.

Later, Erasto gets the opportunity to share his findings with a university professor who visits his school. The professor tries the experiment for himself and discovers that Erasto is right: A hot liquid really does freeze faster than a cold one. How is this possible?

COMMON SENSE DICTATES THAT IT WILL TAKE LONGER FOR HOT WATER TO COOL DOWN TO FREEZING POINT THAN WATER THAT IS ALREADY COLD. SO HOW IS IT POSSIBLE THAT HOT WATER FREEZES FASTER THAN COLD WATER?

The Mpemba Paradox

The counterintuitive finding that a hot liquid freezes faster than a cold one is known as the Mpemba effect or paradox, thanks to Erasto B. Mpemba's work in this field, although the phenomenon was earlier observed by Aristotle and Francis Bacon, among others. The effect does not occur every time, but it has been extensively verified. Denis Osborne, the professor originally approached by Mpemba, suggested that in a hotter liquid there is more convection (where hotter fluid rises to the top and cooler fluid sinks, causing circulating currents), and that this speeds up the diffusion of heat away from the liquid so that it cools faster.

Relaxed and cool

A 2013 study by physicists at the Nanyang Technical University in Singapore offers an alternative explanation, involving the bonds within and between water molecules. In warmer water the molecules move about faster and the distance between them increases (which is why water becomes less dense as it warms up). This reduces the amount of electrostatic repulsion between the slightly charged parts of each molecule, which in turn allows the bonds within each molecule (known as covalent bonds) to relax and contract. This relaxation causes energy to be released, which is the same as cooling, and so warmer water can lose heat energy faster than cooler water.

'My name is Erasto B. Mpemba, and I am going to tell you about my discovery, which was due to misusing a refrigerator . . .'

– **Erasto B. Mpemba (b. 1950)**

006 BROWNED SUGAR

Gaston is making breakfast: cheese toast and coffee with foamed milk. First he puts a slice of bread under the grill. The toast goes brown and releases distinctive odours. When it has gone from light beige to dark brown, he flips it over, puts slices of cheese on it, and puts it back under the grill. The cheese soon melts, and after a while it begins to turn brown in places.

Next, Gaston makes a brew from roasted coffee beans, which are a deep brown and rich in coffee aromas. He has a steam wand attached to his coffee machine for foaming the milk, but he leaves the wand in the milk too long and finds that the milk has also started to turn brown.

WHY DOES ALMOST ANY FOOD SUBJECTED TO SUFFICIENT HEAT, OR EVEN A LOW HEAT OVER A LONG PERIOD OF TIME, TURN SOME SHADE OF BROWN? AND WHY IS THIS ACCOMPANIED BY NEW FLAVOURS AND ODOURS?

Brown is not the only colour

Browning reactions are the primary transformations undergone by food during cooking. They are responsible for much of the look, taste and smell of cooked food. Generally, browning reactions do indeed produce brown-coloured products, but they can also produce a range of colours from yellow, to red, to black.

Caramelisation

One of the simplest browning reactions is caramelisation, which is what happens to sugar when it is heated. Water molecules are driven out of the sugar molecules, which transform into a complex mixture of sugars, sour organic acids such as acetic acid (the main ingredient of vinegar), fruity esters, nutty furans (aromatic hydrocarbons), butterscotch-flavoured diacetyl and brown-coloured polymers. The simple sweet, colourless and odourless sugars are thus transformed into a complex brew of flavours, scents and colours. The appeal of these changes may lie in their similarity to the changes in appearance, scent and flavour associated with ripening fruits, which must have provided a valuable injection of flavour and calories to the diet of prehistoric man.

The Maillard reaction

Alongside caramelisation, the other main cause of browning in foods is the Maillard reaction, named after the French chemist Louis Camille Maillard (1878–1936). In 1912 he discovered that if enough heat is applied, two of the vital components of food – proteins and carbohydrates – will react together to produce distinctive new molecules, including flavour compounds called pyrroles, pyridines, thiophenes and oxazoles. The brown colours themselves come from melanoidins, brown pigments created when amino acids and sugars react with one another during the Maillard process.

'Perhaps cooking with fire was valued by our evolutionary ancestors in part because it transformed blandness into fruitlike richness.'

– Harold McGee, food scientist (b. 1951)

007 BOILING POINT

Theresa's hungry family are clamouring for their lunch. She bustles about the kitchen, preparing a simple pasta dish. Filling a large saucepan with water, she puts it on the stove and turns the heat up to maximum, aiming to achieve a rolling boil, just as her mother taught her. But the large mass of water seems to be taking an age to come to the boil. Initially a fizz of small bubbles forms at the bottom of the pan. After that, however, progress slows. How can she speed things up, she wonders to herself? Then she recalls one of her mother's sayings: 'Pasta should be cooked in water as salty as the Mediterranean' Of course, this was the answer: A generous handful of salt thrown into the pan would bring it to the boil more quickly, wouldn't it?

WE ARE ALWAYS TOLD THAT ADDING SALT TO A PAN OF WATER WILL MAKE IT BOIL FASTER, BUT WHY SHOULD A SPRINKLE OF SALT MAKE ANY DIFFERENCE?

Sea-salty

In fact, the popular wisdom about salt and boiling water has it back to front. Salt actually *raises* the boiling point of water, which means it will take longer to come to the boil. However, because the water will reach a higher temperature, whatever is in the water will be cooked more quickly, and so Theresa is correct in thinking that adding salt will help towards getting lunch on the table as soon as possible. But the effect is small at best: In order to raise the boiling point of 1L (2 pints) of water by just 0.5°C (around 1°F), you would have to add about 24g (0.8 ounces) of salt. This would indeed make the water roughly as saline as the ocean – although the origin of Theresa's family dictum about pasta water more likely relates to taste rather than actual salt content.

Ion mongery

Salt raises the boiling point and lowers the freezing point of water by creating a solution. A salt is a substance that dissolves in water by splitting into charged particles called ions. These ions dilute the water and change the way that the water molecules interact as they change phase from liquid to gas (boiling) or liquid to solid (freezing). Ions also absorb energy without turning into gas. This means that there are fewer water molecules that have enough energy to make the leap from liquid to gas, so that more heat has to be added to make that transition and achieve boiling.

'I know a cure for anything: salt water . . . sweat, or tears, or the sea.'

– Isak Dinesen (Karen Blixen), 'The Deluge at Nordeney' (1934)

008 THE INSATIABLE DUVET COVER

Doug ponders the great questions of life as he does his household chores. Why is there something rather than nothing? Why are we here? Why does my washing machine eat so many socks, but never in pairs?

Thoughts of washing remind him to attend to his next task: the laundry. He strips the bed and puts the duvet cover into the machine along with a bunch of clothes. When the cycle finishes and he takes them out, every single item is lodged inside the duvet cover. The scenario repeats itself when he puts the items in the tumble dryer, but with more annoying consequences: Stuck inside the voluminous cover, the clothes have failed to dry properly. Doug is forced to repeat the process minus the laundry-gobbling duvet cover.

WHY DO ALL THE OTHER ITEMS END UP INSIDE THE DUVET COVER WHEN THE LAUNDRY IS IN THE WASHING MACHINE OR TUMBLE DRYER?

The drunkard's walk

During the washing or drying process some items might enter the mouth of the duvet cover, but surely that probability should be low, with a much greater probability that the clothes would stay outside the cover? The answer may be found in a branch of probability theory known as random walking, which models the behaviour of a system governed by probability. One of the simplest examples is known as the 'drunkard's walk'. This imagines a drunk who staggers out of a bar onto the pavement, and then proceeds to lurch, one step at a time, in one of two directions. At each step he has a 50–50 chance of going east or west. Random walk modelling asks: After any given number of steps, what is the probability of the drunkard ending up back where he started, or at any other given point?

Sock roulette

Applying the drunkard's walk reasoning to the clothes in the tumble dryer, we can imagine that a sock has a 50–50 chance of getting into the duvet cover with each spin of the dryer. If the cover is folded to give multiple turns, we can see that, although the sock has a 50 percent chance (that is, a probability of 0.5) of getting out of the cover at the next turn, it also has a 50 percent chance of getting deeper into the cover. In the latter case, it will now require two 'out' moves to escape the clutches of the cover, and since each of these has a probability of 0.5, there is only a 0.25 probability that the sock will get out, versus a 0.75 probability that it will stay where it was or go deeper into the cover. The more times the dryer tumbles, the greater the chances of the sock being inside rather than outside the duvet cover, and the same reasoning applies to all the other laundry items.

About turn

Another potential factor is that washing machines and tumble dryers often reverse the direction of spin to stop the clothes from squashing up too much, and this sudden shift of momentum helps to force open the mouth of the duvet cover so that it swallows up the other pieces of laundry.

009 RISE AND FALL

Mary is a hardworking, disciplined baker. Paul is a lazy baker. Each of them is making a cake. Mary measures out her flour, eggs, sugar and butter precisely. She warms the butter and beats the eggs before adding them to the mix, and she carefully folds in the flour. She preheats the oven to the correct temperature, and once the cake is inside, she sets a timer and doesn't open the oven until the timer is finished.

Paul just wants to eat cake. He doesn't take the trouble to warm his butter before mixing; he can't be bothered to beat the eggs for very long before adding them to the mix; he whisks the flour into the mix rather than folding it in carefully; and he doesn't measure his ingredients properly. He doesn't check the temperature setting and he opens the oven as soon as the cake starts to smell nice. Paul is disappointed that his cake has not risen, looks like a dinner plate, and feels like a brick, whereas Mary's cake is tall, light and spongy.

NOT UNTIL THE VERY END OF THE BAKING PROCESS WILL THE BAKER KNOW WHETHER THE LABOUR OF CAKE-MAKING HAS PAID OFF. BUT WHAT DETERMINES WHETHER THE CAKE WILL RISE OR FALL?

What Mary knows

Mary's cake turns out well because she adheres to the basic principles of baking, all of which relate to the chemical reactions that must occur in order for the cake to rise. The basic scaffold of a typical cake comprises starch from flour and proteins from egg, but the egg protein must not set (coagulate) too hard and too fast, while the starch must not soak up too much water too quickly before it turns into a sticky gel (this is called gelation). This is why sugar and fats (such as butter) are mixed in, because they help to interrupt and slow these processes. One of the key aims of the mixing process is to aerate the mixture, which means generating loads of tiny bubbles that get captured in the gloopy mix. During baking, the air bubbles need to expand at just the right pace, so that the cake rises evenly. The final stage of the baking process is the hardening of the cake structure as the starch gelates and the protein coagulates, so that the end result is a porous mass that can support its own weight, despite being made up of a network of air-filled cells.

The fall guy

There are multiple opportunities for failure in the cake-making process. When Paul is too lazy to beat his eggs or mix his ingredients for long, he is failing to generate enough bubbles in the batter. By using cold (and therefore hard) butter, he fails to ensure even mixing, while by whisking vigorously, rather than folding in the flour, he causes gluten proteins in the flour to form cross-links, so that the batter is already too rigid to expand properly, even before baking begins. He compounds this problem by making the oven too hot, so that the scaffolding of the cake sets into place before the bubbles can expand, and when he opens the oven he causes sudden cooling of the surface. He probably knocks the cake at the same time, causing air pockets to rupture and collapse.

010 RIPE ON TIME

The banana (together with its close relative the plantain) is the most widely grown and consumed fruit in the world. In some regions of the world people eat hundreds of pounds a year, while globally, on average, annual consumption per person is about 14kg (30 pounds, or roughly 120 bananas). This means that on average each person eats a banana every three days or so. Li consumes her bananas on a less frequent basis: She only eats them on the weekend. But she can only get to the shops on a Monday. In order to avoid her bananas going black and mushy by the end of the week, she always buys green, unripened bananas, but sometimes when Saturday comes her fruit is still green and hard. She has tried placing them in the fridge, but that simply causes the bananas to blacken very rapidly.

Gas control

Bananas are unusual in maintaining a high level of metabolic activity even after they have been picked, which is why they are the only fruit that can be ripened just as well off the tree as on it. This activity – specifically, the action of a variety of enzymes (biological catalysts) – is what causes the ripening, by converting the fruit's high levels of starch into sugar (so that the ratio of starch to sugar goes from 25:1 in green bananas to 1:20 in ripe ones), and by catalysing browning reactions that discolour the skin and flesh. The key to control of banana ripening is a gas called ethylene. This is a simple hydrocarbon that plants use as a hormone (it is also used in industry to make common plastics like PVC). Ethylene can be released by plants in response to injury – it acts as a trigger for defence mechanisms such as dropping damaged leaves – and as a way to control the plant's ripening mechanisms. For instance, bananas have genes that suppress the ripening process; when ethylene receptors on the banana sense the gaseous hormone, they shut down these genes, triggering a cascade of other gene activations and enzyme production that initiates and accelerates ripening. Shipping companies make sure to keep bananas in an ethylene-free environment when transporting them, while retailers will spray their stock with ethylene before putting it on the shelf.

Agent provocateur

Fruit often releases ethylene as it ripens; bananas do this, but to a lesser extent than, say, kiwifruit. Assuming that Li does not have access to a canister of (highly flammable) ethylene, her best bet if she wants to induce ripening at a specific time might be to put the bananas in a bag with a ripe kiwifruit.

'A bad person is like a banana: One alone can turn the whole bunch rotten.'

– African proverb

011 CLEAN SWEEP

Little Bear feasted eagerly on the roasted mammoth flesh at the winter festival, and now his face and hands are greasy with soot and mammoth fat. Grandmother cuffs him on the ear when she sees his grimy face. 'Clean yourself up,' she commands. 'The shaman arrives soon and you cannot come before him looking like a dirty midden hound.'

Little Bear runs to the stream at the edge of the camp and splashes cold water onto his hands and face, but because of the grease it just runs off. He scrubs at his hands with sand but it makes little difference. Disconsolate, he wanders back to the great hearth where the fire is now smoldering and complains to his sister that nothing will shift the dirt. 'I will help you,' she says. 'Go and cut a piece of blubber from the mammoth carcass and bring it back to the fireside.' He does as she says, then looks on as she heats the fatty blubber in a bowl until it melts, and mixes in ash from the fire. As the mixture cools into paste, she shapes it into a cake, which she presses into his hands.'Take this to the hot springs below the caves, and use it to wash yourself in the warm water.'

Little Bear treks up to the springs and finds, to his surprise and delight, that the dirty-looking mixture seems to have magical powers to cut through the greasy dirt on his skin, leaving it clean, though oddly slippery. Now presentable, he returns to the camp and asks his sister, 'What do you call this magic cleaner?' 'We call it soap, but if you want to know how it works, come back and ask me in 10,000 years when they've invented science.'

'What soap is to the body, laughter is to the soul.'

– Yiddish proverb

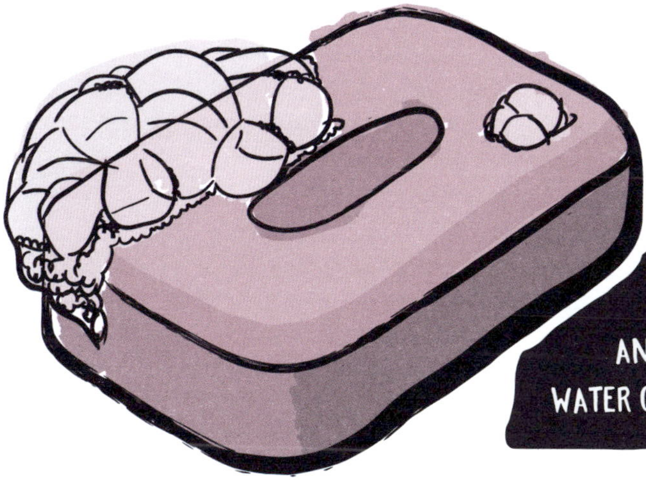

HOW DOES SOAP REMOVE GRIME AND GREASE WHEN WATER CANNOT SHIFT IT?

Heads and tails

Soap works by bridging the divide between two different and mutually repelling worlds: the polar and non-polar. Water and water-soluble substances are polar, which means they have, or can form, electrically charged ends (or poles). Through the attraction between negative and positive poles, these substances can form bonds that make it possible for them to dissolve in water. Hydrocarbon molecules, on the other hand – such as fat, oil and grease – are non-polar, and cannot dissolve in water. This is why oil and water do not mix. It also means that it is hard to wash away a blob of grease with just water, because the water cannot penetrate and lift the grease.

Soap molecules are formed by a reaction between long hydrocarbon molecules – such as fatty acids of the sort found in mammoth blubber – with alkaline salts of sodium or potassium, such as those found in wood ash. The result is a long hydrocarbon tail, which is non-polar and can dissolve in non-polar substances like oil, attached to a polar salt head, which is said to be hydrophilic (water-loving), because it can dissolve in water. Soap can thus bind to oil, breaking it up into tiny droplets surrounded by soap molecules with their hydrophilic heads facing outwards. In this way, each tiny drop repels the other but happily binds with surrounding water molecules to create an emulsion (a suspension of tiny droplets), which is easily washed away.

012 SPARKLING WIN

Juanita and Paco bid farewell to the last of the guests and began the weary task of clearing up. Bottle after bottle goes into the recycling, until Paco holds one up and says, 'This one is less than half-empty; seems a shame to throw it away.' Juanita agrees, but she hates flat champagne. 'How can we get it to keep its sparkle?'

Her mother-in-law pipes up: 'Put a silver teaspoon in the neck of the bottle. My father used that trick for fifty years and he never wasted a drop.' Juanita snorts; it sounds like an old wives' tale. Would her mother-in-law's spoon trick really work?

ONCE YOU'VE OPENED A BOTTLE OF CHAMPAGNE, IS THERE ANY WAY TO STOP IT FROM GOING FLAT?

Spoon trick

Unfortunately, Paco's mother's trick is indeed nothing more than an old wives' tale. Even proponents of the technique disagree on the details. Does the spoon need to be silver? Should the handle of the spoon reach the wine? There is no plausible mechanism by which a spoon in the neck of the bottle could have any effect on the solubility or release of dissolved carbon dioxide, and experiment shows that it makes no difference to whether or not champagne will retain its sparkle.

Why bubbles?

The fizz in champagne comes from the carbon dioxide (CO_2) that dissolves into it during the fermentation process, specifically during its second fermentation in the bottle. As the yeast turns the sugar in the grape juice into alcohol, it releases carbon dioxide, and since the champagne is trapped in a bottle the brew is kept under pressure and the carbon dioxide stays in solution. In fact, about 7.5g (a quarter ounce) of CO_2 (equivalent to 5 litres or 1.3 gallons of gas) is dissolved in an average bottle of champagne.

When the cork is popped and the pressure released, much of this carbon dioxide comes out of the solution, making the champagne fizzy (although only around 20 percent comes out as bubbles). A single flute of champagne might give off a staggering 20 million bubbles, and the rising and bursting of these helps to carry and disperse some of the flavourful and aromatic compounds that make champagne so deliciously distinctive. If Juanita lets the champagne go flat, the very thing that makes it special will be lost, and with it a significant portion of its flavour and smell.

Cold hold

One factor that does govern the solubility of carbon dioxide (i.e. whether it will stay dissolved or come out as a bubble) is temperature: Solubility increases as temperature decreases. So the colder the champagne (so long as it is above freezing), the less CO_2 will come out of solution. This is one reason why champagne should be served very cold (ideally around 5°C /40°F). Juanita should keep the open bottle of champagne in the fridge and turn its temperature setting right down (though not to freezing), and it should retain its sparkle overnight.

013 FAILURE TO LAUNCH

Undaunted by his bad experience with the robot butler (see p. 14), Roger the inventor turns his attention to realising another childhood dream fed by science fiction visions of the future: the flying car.

This time, Roger is more confident of success. All the technology he needs to make the flying car work exists, more or less. There are powerful ducted fans that can provide enough lift for vertical take-off and landing, and minimise the risk posed by the rotor blades by enclosing them in a cylinder (or duct). There are strong, lightweight composites for the airframes. There are efficient, powerful electric motors, and there might even be room in a car-sized vehicle for enough batteries to power them. There are intelligent systems that can make piloting the vehicle relatively easy. And although a flying car would be extremely expensive, there is a ready market of showy billionaires who would love to avoid congestion on the roads.

After years of hard toil, Roger's prototype flying car is ready. He takes it for a spin without incident, but is arrested as soon as he lands. He is taken to court and his flying car is impounded. As he sits in a holding cell, Roger reflects that the future is as far away as ever.

IN THE 1960S, THE CARTOON FAMILY THE JETSONS ZIPPED ABOUT IN A PERSONAL PRIVATE AIRCRAFT, INSPIRING A GENERATION OF KIDS TO GROW UP WONDERING: 'WHY DON'T WE ALL HAVE FLYING CARS?'

Safety first

A true flying car would be a relatively small, road-legal private vehicle capable of taking off and landing in and around spaces shared with conventional automobiles. It would not require a pilot's licence to fly, and certainly would not rely on an airfield to operate. Perhaps the greatest barriers to this dream are legal, not technical. One major difficulty is figuring out how to arrange airspace access directly to and from built-up areas, rather than airports. Safety is the major concern, with the prospect of unskilled pilots on the loose in the air above crowded cities. One solution might be to make flying cars entirely autonomous, but in this case they would face similar problems to self-driving cars.

Worst of both worlds

Teams around the world are near to, or have reached the point of, developing working flying vehicles that fit the profile, but the only flying car it is actually possible to buy is basically a light aeroplane with folding wings, which can only take off and land from an airport, and requires a pilot's licence to operate. It is more expensive to buy one of these than to buy both a light aeroplane and an expensive sports car, each of which offers superior performance in its field.

'I confess that in 1901, I said to my brother Orville that man would not fly for fifty years ... Ever since, I have distrusted myself and avoided all predictions.'

– Wilbur Wright, in a speech to the Aero Club of France, 5 November, 1908

014 BAD NEIGHBOURS

Professor Bailey of the coleopterology department and Professor Otoko of the pestology department compete over everything. When Professor Bailey won a grant from the National Institute for Science, Professor Otoko secured a bigger one. When Professor Otoko took on a new postdoctoral researcher, Professor Bailey took on two. When Bailey was appointed to the Regional Commission on Research, Otoko was elevated to the National Commission on Research.

They even compete when comparing research trivia from their respective fields of expertise. Professor Bailey, who studies beetles and weevils, likes to boast that the beetle is the most ubiquitous animal in terms of proximity to humans. 'In any city in the world,' declares Bailey, 'one is never far away from a beetle.' Professor Otoko snorts. 'Nonsense. And besides, everyone knows that in the city one is never more than 6 feet from a rat.' 'Oh, please,' sneers Bailey, 'Give me a break! You don't seriously believe that, do you? And even if it were true, that's nothing compared to the prevalence and density of beetle habitation. Why, did you know that there are roughly 63 billion beetles in the UK alone? That's over 1,000 for every man, woman and child in the country. Chances are that you are never more than a few inches from a beetle!'

Otoko is not going to stand for that; determined to prove her statistic about rats, she retires to the library to do some research.

'The inference one might draw about the nature of God from a study of his works [is that He has] an inordinate fondness for beetles.'

– J. B. S. Haldane

The rat problem

Unfortunately for Professor Otoko, Professor Bailey is undoubtedly correct. The popular claim that one is never more than 1.8m (6 feet) from a rat is a myth that probably dates back to around 1909, when a book called *The Rat Problem* by W. R. Boelter estimated the number of rats in Britain, based on feedback from farmers, to be about one per cultivated acre (0.4 ha). Since at that time there were roughly 40 million acres (16 million ha) of cultivated farmland in the UK, and the human population of the country was also about 40 million, he concluded that there was one rat per person. Somehow this got translated into an assumption that one is never more than a body length away from a rat. A more recent estimate is that there are around 10 million rats in the UK, and that in a built-up area one is around 50m (164 feet) away from a rat on average. Rat expert Dr Stephen Battersby suggests that in run-down areas one is more likely to be around 3–5m (10–16 feet) from a rat.

None of these figures for rats compare to the prevalence and number of beetles. In terms of the number of species described, beetles account for a quarter of all life forms on the planet; their range is greater than that of rats; and for sheer numbers the beetle population dwarfs the rat population – so the odds are that you are much more likely to be very close to a beetle than to a rat.

015 DIRTY MONEY

Maisie's lemonade stand on the pavement in front of her house is doing a roaring trade. In fact, cups of lemonade are selling so well at 75p a piece that Maisie decides to increase her price to £1 a cup. At the end of the day she goes inside and puts her money box on the kitchen table with great satisfaction. 'Just look at this, Mum', she says, proudly showing a generous collection of coins and notes. Theatrically, she counts the money out, but as she takes a wad of notes and goes to lick her fingers, in order to help her better count, her mother leaps to her feet with a cry of horror. 'Maisie! Stop this instant! Don't you dare lick your fingers after handling all that filthy, dirty money. In fact, I insist you put down the notes and wash your hands this instant!' *Jeez,* Maisie thinks to herself. *What's got into her*?

WHAT SUBSTANCES ARE ON MY BANKNOTE?

Rogues' gallery

Types of bacteria found on banknotes include: *Escherichia coli (E. coli)*, *Vibrio* spp., *Klebsiella* spp. including *K. pneumoniae*, *Serratia* spp., *Enterobacter* spp., *Salmonella* spp., *Acinetobacter* spp., *Enterococcus* spp., *Staphylococcus*, including *S. aureus* and *S. epidermidis*, *Bacillus* spp., *Streptococcus pneumoniae*, *Proteus* spp., *Pseudomonas* spp., including *P. aeruginosa*, *Shigella* spp., *Corynebacterium* spp., *Lactobacillus* spp., *Burkholderia cepacia*, *Micrococcus* spp. and *Alcaligenes* spp.

Laundering required

Maisie's mother has a point: Money really is dirty. Almost all notes and coins are contaminated with narcotics and pathogens (disease-causing organisms). Studies have repeatedly confirmed that over 99 percent of banknotes in general circulation are contaminated with illicit drugs, particularly cocaine, which binds to the green dye used on notes (notably US dollars) and can stay there for months or years. Even though the typical level of drug contamination on an individual banknote is less than 5 nanograms (5×10^{-9}g, less than one ten-thousandth the weight of a typical grain of sand), modern analysis methods are so sensitive that this can easily be detected. What's more, some drugs can persist at levels high enough to cause illness: In 2012, cashiers at a shop in Michigan, the United States, were poisoned by methamphetamine on banknotes they had been handling.

Now wash your hands!

Perhaps inevitably, given that they are among the most handled and passed-around objects in the world, banknotes are also almost universally contaminated with bacteria and other pathogens – potentially more than a household toilet. A 2002 report in the *Southern Medical Journal* found pathogens on 94 percent of dollar bills tested, while New York University's Dirty Money Project discovered DNA from 3,000 different strains of bacteria on dollar bills. The flu virus can reportedly survive up to 17 days on a banknote. Such germs on money have been traced to fecal, oral and vaginal sources.

016 SHOCK HORROR

Luigi and Alessandro live across the hall from one another. Luigi wears polyurethane-soled shoes and has a nylon carpet. He walks with a shuffle. His car has nylon upholstery. He has dehumidifiers in his apartment, and at work his office is air-conditioned. His office is also crowded with glass desks and has PVC floor tiles. Luigi keeps a lucky rabbit's foot, which he likes to rub.

Alessandro wears cotton slippers and has cotton rugs. He picks his feet up when he walks. His car has leather seats. He works in a slightly damp office with wooden floors and concrete walls.

Luigi is plagued by static electric shocks. He gets them when he turns on the tap in his bathroom and when he reaches for the light switch. He gets them when he leaves the car, and when he goes to open a filing cabinet. Alessandro never gets static electric shocks.

Static electricity

Static electricity is electric charge that is not free to move or flow. A person can build up a charge thanks to the ease with which static electricity is generated by friction. Electrons are negatively charged particles normally associated with positively charged atomic nuclei. Through friction between surfaces, electrons can get knocked loose from one surface and accumulate on another. Normally these electrons will be conducted away and the static charge will dissipate, but if they cannot be conducted away the charge may build up faster than it can dissipate, resulting in a significant voltage (the unit of electrical charge). People have accumulated static charges of up to 15,000 volts, and around 5,000 volts is common. If this charge is discharged all at once, a spark results and this causes the sensation of a shock.

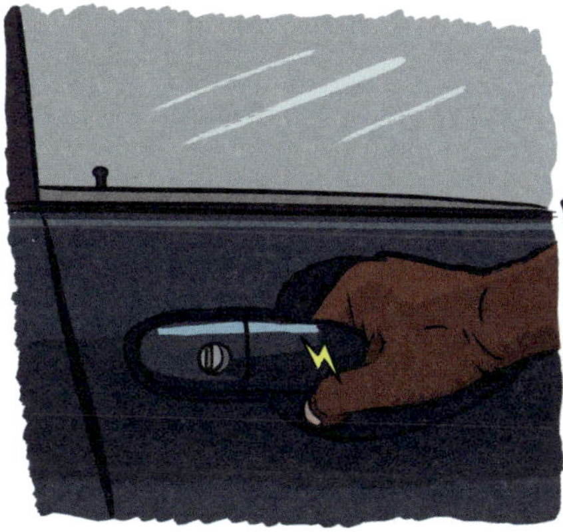

WHY DO WE GET SHOCKS FROM CLOTHING AND CAR DOORS?

Triboelectric series

Some materials lose or pick up electrons more easily than others. Some are much more likely to lose them, and thus become positively charged; others tend to gain them and become negatively charged. Common materials can be ranked into what is known as a triboelectric series, running from most positive to most negative. The further apart in the series two materials are, the more likely they are to interact to generate a static charge.

Shocking behaviour

Wearing or interacting with materials from extreme ends of the triboelectric series increases the risk of generating static charge, as do behaviours that increase friction, such as brushing against things or shuffling your feet. Dry air increases the risk as well. In a car, friction with the seats can generate a charge differential between you and the car, which does not dissipate because the car's rubber tires insulate it from the ground. When you get out of a car, taking your half of the charge with you, touching the metal of the exterior allows the charges to recombine suddenly, causing a shock. You can dissipate this by touching the windscreen (glass conducts the charge away, but slowly), or by touching the outside of the car with your keys rather than your fingers.

017 ASSORTED NUTS

Sarah likes to do good deeds wherever she can. Sitting in the cinema, she notices that the guy in the seat next to her keeps crying out in dismay as he ploughs through his popcorn. He explains that unpopped kernels are getting into each mouthful of popcorn, and they are practically breaking his teeth. Sarah takes the carton from him and shakes it vigorously for a couple of minutes before handing it back. Her fellow cinema patron is delighted; now he can enjoy stress-free mouthfuls of the salty snack.

After the film, Sarah goes to a coffee shop. The barista behind the counter is in despair. The grinder has broken down mid-grind and not all the beans have been ground. The barista complains that she cannot use a mixture of grounds and whole beans in the brewing machine without breaking it. Sarah takes the container of grounds and shakes it vigorously for a couple of minutes before handing it back. The barista is delighted, as she is able to scoop out all the whole beans from the top of the container.

Later, Sarah goes to a party. The host is anxious about the party snacks. One of his guests only likes Brazil nuts, and the host does not want him to have to sort through less-favoured peanuts and cashews in order to reach his preferred snack nut. Sarah takes the container of nuts and shakes it vigorously for a couple of minutes before handing it back. The host is delighted, for now the picky guest will have instant access to the large nuts he most desires.

IN A BOWL OF NUTS, WHY ARE THE BRAZIL NUTS ALWAYS SITTING ON THE TOP?

Granular convection

What is Sarah's secret? Her trick works because of granular convection, a process that governs the motion of particles when agitated. A container full of grains of identical sand, agitated up and down, will display convection currents, with the grains rising up in the centre of the container and falling back down to the bottom around the sides. If the grains are of different sizes, they will assort out, but in a rather counterintuitive way, nicknamed the Brazil nut effect after the tendency of a bowl of assorted nuts to sort themselves into size order, with the largest nuts – the Brazils – at the top. The Brazil nut effect applies to all sorts of different particles, including coffee beans in coffee grounds and unpopped kernels in popcorn.

Trickle down

The primary explanation for the Brazil nut effect is that while the large nuts are able to rise to the top along with all the other particles, they are too large to fit down the narrower spaces around the sides of the container. Instead, smaller particles filter down through these restricted spaces; the smaller the spaces, the smaller the particles must be to filter down through them, so that the smallest ones end up at the bottom. The phenomenon has many complexities: It depends on factors such as density, gravity and air pressure, and it can be reversed by using a conical container. So, if you don't want your Brazil nuts to rise to the top, put your assorted nuts in a cone.

BEING HUMAN

'A thorough study of Human Physiology is, in itself, an education broader and more comprehensive than much that passes under that name. There is no side of the intellect which it does not call into play, no region of human knowledge into which either its roots, or its branches, do not extend.'

– Thomas Henry Huxley, 'Universities: Actual and Ideal' (1874), in *Collected Essays* (1893)

018 SWEATY BEDDIES

The children whooped and shouted as the ice man wheeled his little cart into the back garden and levered the great block of ice out onto the lawn. It stood there, glistening wetly in the summer heat as clouds of mist fell from its sides like a mini-Niagara Falls. The kids crowded around, tentatively reaching out their little hands to touch its sides and screaming with delight at how cold it was.

'But, won't it melt?' inquired Hilda, the youngest. She pointed up at the blazing sun: 'It's really hot'. He ruffled her hair fondly. 'It will melt eventually, but it'll take a long time, sweetie.' 'But when Billy dropped his ice lolly on the pavement yesterday it turned into a puddle really quickly,' Hilda objected. 'Well, when a block of ice is as big as this, there's a lot more inside than outside, if you see what I mean, so it won't heat up nearly as fast as Billy's ice lolly.'

Hilda didn't really see what he meant, and she was still asking about it at bedtime. 'Hush, now, and go to sleep,' said her mum, and she drew up the covers because there was now a chill in the air. When she checked on the little girl two hours later, she saw that her brow was wet with sweat, and when she went to smooth away the curls stuck to Hilda's brow, the pillow was sopping wet too.

WHY DO KIDS SWEAT SO MUCH WHEN THEY SLEEP? IS IT NORMAL FOR MY KIDS TO BE BATHED IN SWEAT AT NIGHT, OR SHOULD I BE WORRIED?

Size matters

Like adults, children may sweat at night for a variety of reasons, most notably when it's hot and when they are sick. But profuse sweating is common and normal in healthy children, even when the ambient temperature is low. The basic reason is that children have a high ratio of surface area to volume, compared to adults (three times more for a baby, and about 65 percent more for toddlers). As the size of an object increases, its volume increases to the power three, but its surface area is merely squared, so that an elephant, for example, despite having much more surface area than a mouse, has a much lower surface area to volume ratio. Since heat is lost – or, when the ambient temperature is above that of the body, gained – through the surface, children will tend to gain heat much more quickly relative to their size than an adult (just as they will also lose it more quickly). Coupled with the fact that temperature regulation responses in the body do not mature until late teenage years, a child's core temperature will increase quickly when the room is hot or when they are under the covers, and this leads to sweating.

Gland ahoy

Although they actually sweat less per unit area of skin, and produce much less sweat per sweat gland, compared to adults children have a higher density of heat-activated sweat glands (partly due to having a similar number crammed into a smaller area). They also spend more time in deep sleep (as opposed to REM [rapid eye movement] sleep), which may also be associated with higher levels of sweating.

019 GREY MATTERS

Natalie had a new customer at the hair salon. Mr Aziz insisted that grey hair was the 'in thing' and wanted to know how he could achieve that genuine salt-and-pepper look. The simplest way, Natalie told him, would be to wait. His chances of going grey, like those of most people, would increase by 10 to 20 percent every decade after the age of 30. Mr Aziz protested that he was already 50 years old. Natalie pointed out that the general rule of thumb among hairdressers is that by age 50, half of people have lost the colour in half of their hair, but that this rule mainly applied to Caucasians. An academic study had found that 74 percent of people aged between 45 and 65 had grey hair, with an average intensity of 27 percent grey.

So, Mr Aziz was among the minority 26 percent of 45- to 65-year-olds without grey hair. Why, he wanted to know, was he not going grey? And why *were* most of his contemporaries?

Natalie considered: Did he want the proximate or the ultimate answer? Mr Aziz wanted to know the difference. 'Well,' said Natalie, 'the proximate reason is the first-hand cause – what happens at the cellular and molecular level with melanin and hair follicles – while the ultimate explanation would relate to the evolutionary rationale for greying of hair.' Mr Aziz was not used to this kind of chat from his barber, but he encouraged her to give him both proximate and ultimate answers to the question.

Peroxide greys

Hair colour comes from various types of melanin pigment that are present in the keratin protein that makes up strands of hair. As the keratin is produced in the hair follicle (the structure from which an individual hair grows), melanin is injected into it by a cell called a melanocyte, positioned at the tip of the follicle. As the amount of melanin added to the keratin declines, the colour begins to fade, first becoming grey and eventually, when there is no melanin at all, white. Melanin production declines for various reasons, some external (i.e. environmental, such as exposure to pollution) and some internal (such as genes and hormones). According to a 2009 study, one reason that melanin dries up is that the melanocytes suffer 'massive epidermal oxidative stress', which is where a buildup of naturally occurring hydrogen peroxide both interferes with the pathway for production of melanin and damages the mechanism for repairing that pathway.

Ladies prefer greys?

The ultimate evolutionary rationale for grey hair presumably lies in some selective advantage conferred by the change. The simplest explanation would be that making melanin for hair costs the body resources in material and energy, so stopping making it frees up those resources for survival. It may be that while early humans were still fertile and competing for mates for procreation, it was worthwhile expending resources on darker hair that might act as a signal of fitness.

An alternative explanation is that grey hair, through its association with ageing, may have signalled valued attributes such as wisdom and experience – although there is no evidence to support this notion and many biologists deride it as 'just-so' storytelling.

020 BITE ME

Nyanga slapped the back of her leg, again. The repellent wasn't working. She looked across to her twin sister, who was sleeping on the far side of the tent. Lupita looked serene – as well she might, the lucky thing. Nyanga would be serene, too, if it weren't for these infernal mosquitoes. She got out of her sleeping bag and went to sit by the campfire. She found several other members of the tour group already sitting there. 'Couldn't sleep either, huh? Was it the mosquitoes?' They all nodded ruefully, pointing to the many bite marks on ankles, wrists, calves, necks and almost every other body part.

Nyanga surveyed her fellow victims, who constituted about a fifth of the tour group. There was a young guy swigging a beer; a large, heavy man; a pregnant lady; a guy in a bright red T-shirt; and a lady whom she had seen jogging just before bedtime. 'Why us?' complained the fat guy. The pregnant lady said that she'd read that about 20 percent of people were highly attractive to mosquitoes, and that there must be something about them that made them smell particularly tasty.

'Actually,' piped up the young guy with the beer, 'they're not eating your blood – it's the female mosquitoes that bite you, and they want the blood as part of their reproductive cycle. Mosquitoes are vegetarians, believe it or not.' Nyanga already knew this – she was a biologist specialising in the habits and life cycle of *Aedes aegypti*, the mosquito that spreads dengue fever and Zika virus. But what really annoyed her was that twin studies had shown that, when it comes to the difference between people in their attractiveness to mosquitoes, over 67 percent of that difference is down to their genes. In other words, being a mosquito target is at least 67 percent heritable. Yet her identical twin sister, against all the odds, was right now sleeping like a baby, apparently untouched by the pesky bugs.

Red flags

Many different chemical signals are given off by the human body: Some it makes itself, such as pheromones or sweat; others are made by the vast population of microbes that lives on humans. Around 400 different chemicals are currently under investigation over their role in attracting mosquitoes, but scientists already know about some major 'pull factors':

- Carbon dioxide (CO_2): Humans exhale this gas with every breath, and mosquitoes use this to detect likely targets from some distance – up to 50m (165 feet) away. The bigger and more metabolically active you are, the more CO_2 you will generate, so this underlies other 'attractiveness factors' – for instance, overweight and/or big people will attract more mosquitoes, and this is why adults get bitten more than children.
- Substances in sweat, such as lactic acid, attract mosquitoes, so exercising will make you more of a target.
- Pregnant women, due to their size and weight, exhale about 21 percent more CO_2 and are on average 0.7°C (33.3°F) warmer than other people, making them more of a target.
- Drinking one bottle of beer makes you more attractive to mosquitoes (nobody knows why).
- Mosquitoes use vision to guide them at close range, and find it easier to see distinctive colours, such as bright red or black clothing.
- People with blood type O are twice as likely to be bitten as those with type A, while those with type B come in the middle.

021 RUN, DON'T WALK

At a forthcoming Olympic Games there is controversy as a new race is introduced. The Flats and Hills race is a combination of steeplechase, marathon and hill-running. The course is extremely long, so the race is an endurance test, but it also passes over several different types of terrain. It begins with several laps of the track before the course leaves the stadium and starts to climb up and down a series of hills, crossing broken ground. Then there is another flat section, followed by more hills, before a final lap of the stadium track.

Having never before competed in a race like this, the athletes are uncertain of the best strategy. Most of the distance runners plan to stick to what they know best, and decide to run the entire length of the course, even though it is longer than a marathon. Most of the distance walkers opt to stick to what they know best, which is walking without breaking into a run. A few athletes, however, plan to adopt a mixed strategy, running the flat sections and walking in the hills. Who will win?

IS RUNNING ALWAYS FASTER THAN WALKING? THE ANSWER MIGHT SEEM SELF-EVIDENT, BUT IN FACT THE BALANCE BETWEEN SPEED AND EFFICIENCY IS COMPLEX AND CONTINGENT.

Mind the gait!

The two primary gaits employed by humans are walking and running. During walking, the knees are locked so the legs stay stiff, which means that they act like inverted pendulums. A pendulum conserves the energy of its upswing as potential energy (produced by the force of gravity acting on the raised pendulum), and converts it back into kinetic energy on the downswing with a very high degree of efficiency. This makes a walking gait very efficient in the right circumstances. In running, however, the legs act more like pogo sticks, bouncing along, which can also be a highly efficient mode of travel. In running, the muscles and tendons of the legs act like springs, storing the energy generated when the foot impacts the ground and returning about 95 percent of it to propel the body forwards.

Energy saver

Humans can run slowly and walk fast (up to 4.6m [15 feet] per second using a speed walker's gait in which the hip is dropped at each step), so speed is not necessarily the determining factor in which gait to adopt. What the body seeks is the greatest efficiency in terms of energy expended for speed generated. Above a threshold speed (which will depend on the terrain and gradient), it becomes more efficient to run than walk. What this means for the Flats and Hills race is that athletes who adopt a mixed strategy, running on the flat and walking in the hills, will achieve the optimum return of speed for the energy they invest. An athlete with the stamina to run the entire course might well be fastest, but over a long enough course the energy demands for constant running will be too great and the athlete with the mixed strategy will triumph in the end.

022 ARMY OF THE DEAD

President Ryan turned away from the window. 'Heaven help us, there are more of those cursed . . . things out there than ever.' His chief of staff snorted: 'Why don't you call them what they are, sir? We all know . . . they're zombies, sir, the living dead!' Ryan crossed himself and muttered despondently, 'When there's no more room in Hell, the dead will walk the Earth.' 'With all due respect, Mr President,' barked Admiral Lin, 'I'm not ready to throw in the towel just yet. All of humanity is united behind us, along with the military might of every nation on Earth. What we need to figure out is exactly what we are up against.'

She turned to Professor Patel. 'Assuming for the moment that the Vatican statement is correct, and that the dead really are returning to feed on the brains of the living, what does this mean in terms of numbers? There are more than 7 billion people on the planet. That, coupled with the exponential growth rate of the global population, must surely mean that we outnumber these goddamn zombies?'

Professor Patel shifted uneasily. 'I wouldn't be so sure, Admiral. I've been doing a back-of-the-envelope calculation, and the numbers don't look good . . .'

ARE THERE MORE PEOPLE ALIVE TODAY THAN HAVE EVER LIVED IN THE HISTORY OF MANKIND?

Arithmetic vs. exponential

Arithmetic growth is when a population increases by steady increments, for instance by adding roughly the same number of people every generation. So, if the original population is n, then at the next generation it will be $n + n$, then $n + n + n$. After, say, 6 generations, the population will be $6n$. For instance, if the original population was 10, after six generations the population will be 60.

Exponential growth is when a population increases by multiples, which is to say by increasing the exponent or power to which the original number is raised. If the original population is n^1, at the next generation it will be n^2, then n^3, and so on. After six generations, the population in this scenario would be 10^6, or 1,000,000 (1 million). In the case of exponential growth, each generation is an order of magnitude larger than the one that went before, so that the million members of the sixth generation would indeed outnumber the 111,110 people in the preceding five generations by nine to one.

Although global population growth has accelerated sharply since the Agricultural and Industrial Revolutions of the eighteenth and nineteenth centuries, and even more quickly since the twentieth century, growth has been nowhere near exponential. Indeed, for most of human history global population has increased very slowly, if at all. On this basis alone, it is highly unlikely that the living outnumber the dead.

Zombie apocalypse

In 2011, according to the UN, the global population passed 7 billion. But according to the Washington-based Population Reference Bureau (PRB), the total number of people who have ever lived is estimated to be around 107 billion, meaning that come the zombie apocalypse, the dead will outnumber the living by approximately 15:1. In fact, the PRB admits that its figure may be an underestimate, since very high rates of infant mortality through most of human history likely mean that birth rates were also very high, and that the ranks of the undead would be swelled by many who died young.

023 BREAD BASKET CASES

Peter surveyed the sign installed above the frontage of his new shop with satisfaction: 'Peter Pain – Artisan Breadmaker.' Alice from the health-food shop next door, 'Eco Eats', came out to welcome him to the street. She looked regretful when she saw the sign. 'I'm sorry to tell you that people around here are very anti-gluten.' Peter protested: 'But gluten is a harmless and exceedingly remarkable natural substance,' he told her. 'The largest composite protein molecule in the world, its incredible elastic and plastic properties make possible the myriad forms and features of bread, pastry, pasta and many other products.'

Alice handed Peter a leaflet that set out in depressing detail the rise of gluten-free food in major markets. Peter was aghast; he couldn't understand what people could have against a substance that had been a dietary mainstay for millennia. Was it really possible that, in the space of a decade or so, a huge chunk of the population had developed a hitherto unknown, life-changing dietary disorder? Or was it all just a fad?

IT SEEMS LIKE EVERY OTHER PERSON TODAY CLAIMS TO BE GLUTEN INTOLERANT. BUT WHAT IS THE REAL INCIDENCE OF THIS CONDITION?

The march of gluten-free

In the United States, in the five years to 2014, sales of gluten-free foods grew 34 percent annually, reaching $973 million, and they are projected to reach $2.34 billion in 2019 – a 140 percent increase from 2014. In 2015, nearly a quarter of all product launches in this sector were claimed to be gluten-free.

Sensitive subject

Why would people abandon delicious, cheap, nutritious gluten-containing foods in droves? The answer is that there has been an explosion in the incidence of both celiac disease – a gluten-induced inflammatory disorder of the lining of the small intestine – and non-celiac gluten sensitivity (NCGS), also known as gluten intolerance, which can cause a range of chronic symptoms, including irritable bowel syndrome, chronic fatigue, eczema, depression and even tooth decay. According to some estimates, the incidence of NCGS has gone up from 1 in 2,500 to 1 in 133 between 1999 and 2009, and it is even claimed that up to 70 percent of the population has latent gluten intolerance, with 30 to 40 percent of this group going on to develop NCGS.

Rising (glu)tension

NCGS is generally self-diagnosed, or depends on clinical rather than diagnostic markers – that is, on symptoms reported by the patient, rather than something that can be measured such as elevated levels of particular substances in the blood. A widely reported 2013 study from Monash University in Melbourne, Australia, on people who had self-diagnosed as gluten intolerant, found 'no evidence of specific or dose-dependent effects of gluten in patients with NCGS'.

On the other hand, it is known that modern wheat varieties and industrial bread production have dramatically increased levels of gluten in most packaged bread. Peter's handmade products, with longer fermentation, may well be less problematic for gluten-intolerant people.

024 GET A SHOT, OR NOT?

Huma has received a letter from her doctor telling her to bring her baby into the clinic to receive a set of immunisations, including ones for tuberculosis and hepatitis, and combined ones for measles, mumps and rubella (MMR), and for diphtheria, pertussis (whooping cough) and tetanus. But when she compares notes with the other parents at the school gates, she hears a whole range of horror stories.

One woman says that vaccines are made with mercury, a poisonous metal, and someone else says that they also contain formaldehyde, the chemical used to preserve dead bodies at the mortuary. One mother claims it is some kind of government conspiracy, because why else would they make you get vaccinated for measles, when no one has died of measles since the year 2000? One of the dads has heard that several kids have died from getting the measles vaccine, and asks how it makes sense to risk killing a baby in order to 'protect' it against a disease that no longer exists. Another dad relates a tale about a kid that got the MMR vaccine and then developed autism, and everyone agrees that they have heard that the vaccine causes autism.

Now Huma doesn't know what to do. Should she get her baby immunised? Wouldn't it be safer not to?

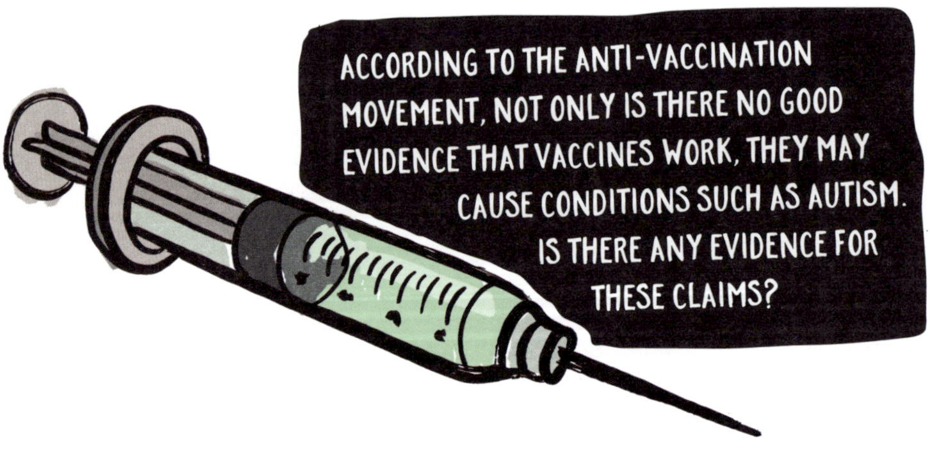

How vaccines work

The human immune system can acquire immunity to many disease organisms by generating antibodies to them on first encounter and then 'remembering' how to make those antibodies, so they can be deployed immediately the next time the germ is encountered. Vaccines make use of this ability by exposing the immune system to very weakened, almost dead forms of a germ, or to small parts of germs. In this way, the vaccinated person can acquire immunity without ever having to catch the disease.

Vaccine myths

Huma's school gate informants are peddling some of the most widespread and pernicious myths about vaccines.

- Vaccines have historically been made using mercury and formaldehyde at different stages of manufacture, but even when these are still used, the minute traces present in vaccines are far smaller than are encountered naturally. For instance, formaldehyde naturally occurs in apples and pears at much higher levels than in vaccines.
- A non-toxic form of mercury was used in a preservative associated with vaccines up to the start of the twenty-first century, at amounts lower than the more toxic form of mercury found in a can of tuna, and there is no evidence of any harm as a result. Nevertheless, to assuage public fears, this preservative is no longer used in vaccines for children.
- Children have died of measles in the developed world in recent years. Before this, measles deaths had declined to zero in most developed countries following the introduction of vaccines. There have been no deaths due to measles vaccines.
- Vaccines can cause side effects, like any medicine, but there is no evidence from valid research linking vaccines to autism, while a wide range of studies involving millions of children shows that there is no link. The original research that claimed there was a link has been completely discredited and repudiated by the journal that published it.

025 HAIR TODAY . . .

Imagine a world in which a shiny, bald pate is a sign of youthful vigour, and a hairy scalp is a dreaded harbinger of approaching old age. Our attitude to hair might be rather different – and yet, perhaps not all that different . . .

Shen could not stop looking at the advertisement for the hair clinic. The treatments were surprisingly affordable. Now that the kids had left home and he had some money to spare, why shouldn't he spend it on fixing his hair problem?

He knew what people thought when they saw his luxuriant crop of thick, shaggy hair. Why should he have to put up with every man's pitying look, and every woman's stifled giggle? It was so unfair; until the age of 40 he had been as smooth-pated as the next guy, but then the stray wisps had begun to appear around the sides, the telltale fuzz creeping across his scalp until there wasn't a bare patch to be seen. He blamed his grandfather, who had gone hairy at the age of 21. With genes like that, what chance did he have? Shen remembered his own father, always so proud of his gleaming dome: How he had encouraged the infant Shen to oil it for him, to marvel at its shininess, and how mortified he had been when the first stubble appeared.

Well, enough was enough: Shen would book an appointment today, and within the week he'd be as bald as a teenager! His wife would look at him with new eyes; his boss would give him that promotion he deserved.

Only it didn't quite work out that way. The hair-removal treatment was a complete success – but now that he looked ten years younger, his boss took him less seriously than ever. He suspected his wife had started an affair with the ponytailed guy at the end of the street, and as for the girl at the tennis club – he had overheard her saying that she was only interested in 'distinguished-looking grown-ups, not childish losers'. Perhaps growing old gracefully might have been the best policy after all.

The bald facts

In Shen's world the normal pattern of male hair growth is reversed, but the consequences seem to be much the same. In our world one man in six goes bald at some point, while one man in 20 has a receding hairline by the time of his 21st birthday. Since baldness seems to be a heritable characteristic, it must have a genetic basis, which in turn suggests that the genetic predisposition for baldness arose from, and is maintained by, some selective pressure. One hypothesis is that for our hominid ancestors, baldness signalled maturity and the command of resources associated with age, making bald hominids more attractive as potential mates. Supporting this theory is the fact that in chimpanzees such an association is indeed observed between social status and baldness. Bald chimpanzees tend to be respected older males, more likely to sire offspring.

026 THE GRANDMOTHER HYPOTHESIS

'Ugh!' Sharp Spear turned away from the shaman and looked out from the rock shelter. 'You have got to be kidding me . . . My mother-in-law's having another baby? But she's more than 60 winters old. By the Great Mother, I've got six of my own to take care of, but do you think their grandmother has time to help out once in a while?' 'Let us consult the Great Mother,' suggested the shaman.

They approached the massive stone idol, showering it with the ritual concoction. The air shimmered, and the Great Mother materialised. 'O Great One,' intoned Sharp Spear, 'please can you change the fabric of our beings so that my mother-in-law will stop squeezing out babies and help out more with her grandkids?' The Great Mother considered, then replied, 'But her grandchildren carry just a quarter of her bloodline, while her own children carry half. In terms of perpetuating her bloodline, she gets a better return on the resources of food, energy and time that she must expend in caring for her own children rather than yours.'

'However,' objected the shaman, 'at her age her seed is diminished in its vital force, and her own body is so worn out that she runs a greater risk of perishing in childbirth. Surely the wiser investment is in her offspring's offspring?' 'Good point,' conceded the Great Mother. 'Henceforth, women shall suffer mood swings, hot flushes, insomnia and many other woes as they lose fertility in middle age.' 'Ugh,' thought Sharp Spear. 'This may have been a mistake . . .'

The mathematics of genetic profit

Menopause seems to fly in the face of evolutionary common sense: Somehow genes have been selected that limit their own chance of being passed on, by ending fertility well before the end of their bearer's life span. The traditional explanations are that menopause is an inevitable side effect of ageing or a trade-off for improved fertility in earlier life; or that it evolved so that older women could invest resources in their grandchildren (known as the 'grandmother hypothesis'). But as the Great Mother points out, the maths of this doesn't necessarily work out: only 25 percent of a woman's genome is present in her grandkids, versus 50 percent in her own children.

One possible explanation is that, as in chimpanzee groups, in early human societies women went to live with their partners' social groups – known as female-bias dispersal – so that in a social group the younger child-bearing women would share no genetic relationship with the older child-bearing women (their mothers-in-law). Thus, the younger women would have gained no 'genetic profit' by helping their mothers-in-law to raise babies (who would nominally be their brothers or sisters-in-law but would have no genes in common with the younger women). On the other hand, the mothers-in-law would gain a genetic profit from investing in their daughters-in-law's offspring, since these would be their grandchildren bearing 25 percent of their genes. In the tension over resources to support fertility, the 'economics' of genetic profit thus mitigated against continued fertility for the older women in the group, since on balance the most mutually beneficial way of expending resources was in support of the fertility of the younger women.

Menopause in other species

Elephants have babies in their 60s, and some whales give birth in their 80s, so there is not necessarily an inevitable link between age and fertility. The only other animals known to undergo menopause are killer whales and their relatives the pilot whales. In these species, individuals live in tight-knit family groups, and postmenopausal females do indeed seem to help care for their grandchildren. They may spend a greater propoertion of their lives postmenopausal than humans do.

027 GASPING FOR AIR

After decades of painstaking research, endless prototypes and relentless derision from the academic community, Maria was finally ready to try out her new invention. Never before had stem-cell seeding and genctic manipulation been used in tandem in this way, to achieve the ultimate body modification, the bio-hack Maria had dreamed of ever since she was a girl. She still remembered the first time she had dived into the crystal-clear water, how she had marvelled at the colours and forms of the creatures on the reef, and the bitter sense of injustice she had felt when the burning in her lungs compelled her to return to the surface. Now, at last, Maria would fulfill her lifelong ambition and swim with the fish on equal terms.

FISH USE THEIR GILLS TO EXTRACT OXYGEN FROM THE WATER; WHY CAN'T HUMANS USE THEIR LUNGS TO DO THE SAME?

Long years of study in biology had taught her the nature of the problem: The gills of fish are perfectly adapted to extract the oxygen dissolved in seawater, just as human lungs are adapted to extract oxygen from the air. In fact, the two systems are fundamentally alike, since mammalian lungs depend on creating sea-like conditions within themselves: Air drawn into the lungs dissolves into the mucus that lines them, and, just as in the gills of a fish, blood vessels crowd close to the surface of the mucous membrane, enabling dissolved oxygen to diffuse across the membrane and into the blood. A few tweaks in the human genome could create gill-like structures within human lungs that could withstand the rigours of exposure to seawater. By injecting herself with pulmonary stem cells carrying the genetic modification, Maria had succeeded in doing what everyone had said would be impossible: She could breathe water!

Plunging into the ocean, Maria swam to the seabed and took her first breath. She felt the cool liquid rush into her modified lungs, and effortlessly expelled her first lungful of seawater, drawing in another. She was breathing underwater! Quickly, however, she began to feel light-headed, her senses dimmed, her concentration wavered. Something was very wrong. As consciousness faded, Maria realised she had overlooked one simple fact.

The air down there

Maria remembered too late that although there is indeed oxygen in water – which we could theoretically extract if we had suitable biological equipment – there simply isn't enough of it. Seawater contains 1.5–2.5 percent dissolved air, and about a third of this is oxygen – enough for cold-blooded fish, with their relatively low metabolic rates. But mammals evolved to breathe in a medium 40 times richer in oxygen, enabling them to support warm-blooded, high-intensity metabolisms. In order for her brain and body to function, Maria needs access to air, which is 21 percent oxygen, rather than water, which is just 0.5–0.85 percent oxygen. The finest gills in evolutionary history cannot make up for this fundamental disparity, which is why mammals that have returned to the sea, such as seals and cetaceans (whales and their relatives), have kept their lungs and still need to come to the surface to breathe air.

028 DESIGN FLAW

Enki, the Sumerian god of creation, had assembled a group of good and princely lesser gods to help create man, a new technology needed to serve the gods and maintain the order of the cosmos. They gathered in the briefing room, sipping chai lattes, and Enki started scribbling furiously on a whiteboard. 'Primarily,' he told them, 'this is an engineering problem. We've got to create functional sentient life-forms using some pretty basic materials, and above all, people, we have *limited resources*. I cannot stress this enough: Don't just throw anything into the mix – it must be viable, economic, physiologically affordable. OK, Anzu? No wings or flying, alright? Now, let's box this out.' Split into working parties, the good and princely gods started to tackle the many challenges of designing a functional species.

MALE NIPPLES SEEM TO SERVE NO BIOLOGICAL PURPOSE, BUT THEY CAN CAUSE HEALTH PROBLEMS SUCH AS BREAST CANCER. WHY HAVE THEY NOT BEEN LOST DURING EVOLUTION?

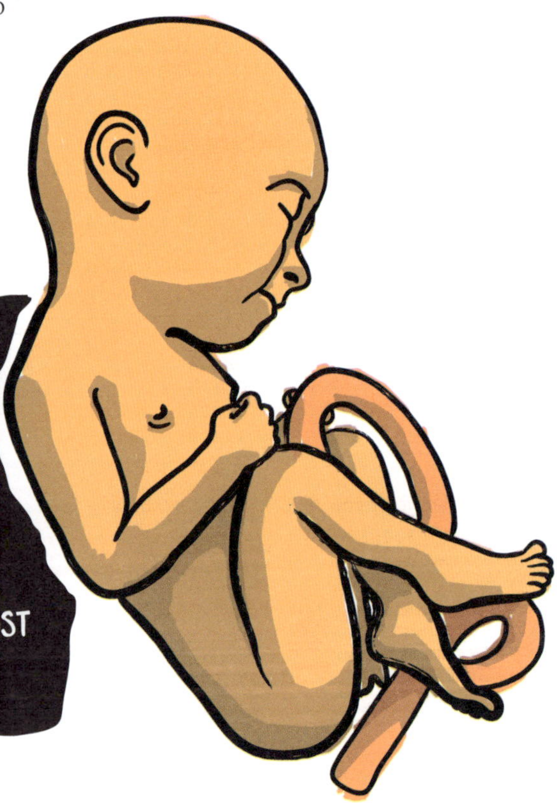

The reproduction team quickly decided to go for a binary sexual model, the most straightforward way of assuring healthy genetic variation. The development team, tasked with achieving gender differentiation, went for a basic template generated at fertilisation, which could be acted upon by sex hormones from week seven of gestation (Ninhursag, leading the pregnancy and birth team, had pushed for a long, 40-week gestation).

Presenting to Enki, they explained the process: 'A lot of the basic body structures that we'll see in the adult female are laid down first irrespective of gender, and then circa week seven, if the Y chromosome is present in place of one of the X chromosomes, the male sex hormones kick in and the labia fuse into a scrotum, while the clitoris becomes the penis.' 'Hmm,' reflected the creator god, 'seems a bit over-engineered, but OK. What about this feature up here, though, what's this?'

The lesser gods exchanged guilty looks. 'Um, those are nipples. The female will need those for feeding her infants.' 'But you have them developing circa week four, and look, over here, your boy babies have still got them after gender differentiation. So basically, the males will be stuck with redundant features?' 'Um, well, perhaps we can iron out the kinks in version 2.0?'

Not worth losing

The design set out by Enki and the other gods is indeed how sexual characteristics develop in the womb. Fetuses of both sexes start off with nipples, which develop around week four of gestation, but while the genitals change in the presence of sex hormones, the nipples do not. Normally, evolution favours the loss of redundant features, since they cost energy to make and maintain, and this is probably why the males of other mammalian species, such as horses, rats and mice, have lost their nipples. But male humans retain theirs, although they serve no function. Presumably the physiological cost of nipples is relatively trivial, and selective pressures against them simply too weak. So, the short answer to the question, 'Why do men have nipples?' is that all fetuses have them – and all fetuses have them because the female ones will need them later.

029 THE EDGE OF SURVIVAL

Spielman's car has broken down in the middle of the desert and he is a few days' walk from the nearest town. On these desert roads you can wait a week before another car comes by. Worst of all, Spielman has not brought any water. Just how long, he wonders, will he be able to survive out here without water?

After a day, Spielman's saliva is thick and rank, and his tongue sticks to his teeth and the roof of his mouth. There is a lump in his throat, which he tries to dislodge by swallowing repeatedly, but to no avail. By the end of the second day he experiences excruciating pain in his head and neck. The skin shrinks on his face. Overnight he begins to hallucinate. By the third day his tongue has become a dead weight, knocking against his teeth. His body is mummifying, even though he is still alive – barely. His eyelids crack and his eyes weep tears of blood. His throat has swollen so much he can hardly breathe and he believes he is drowning. When the transport police find him that evening he looks and acts like a zombie; he is on the very edge of death.

EVERYONE KNOWS HOW IMPORTANT WATER IS TO LIFE, BUT WHAT IS THE ABSOLUTE LIMIT OF HUMAN ENDURANCE OF THIRST AND DEHYDRATION?

The living dead

Spielman's ordeal is based on the true story of a man lost in Arizona's Sonoran Desert in 1905. Amid the savage desert heat, W. J. McGee, Director of the St. Louis Public Museum, saw a man crawling along a desert trail, reduced to a state of living mummification. 'His lips had disappeared as if amputated . . . his teeth and gums projected like those of a skinned animal, but the flesh was black and dry . . . his skin [had] turned a ghastly purplish yet ashen grey . . . We found him deaf to all but loud sounds, and so blind as to distinguish nothing save light and dark . . .' The man was Pablo Valencia, a prospector who had ventured into the desert with insufficient water.

Stages of thirst

Incredibly, McGee was able to nurse Valencia back to life, and later wrote a paper on 'Desert Thirst as a Disease'. In it he describes successively more appalling effects of thirst and dehydration: the cotton-mouth phase; the compulsive swallowing; hallucinations; drying out of the tongue, blood sweats and tears; throat swelling and delusions of drowning; and finally the 'living death' stage. This was the condition in which McGee found Valencia, after the errant prospector had survived for six-and-a-half days in the burning desert heat with no moisture but his own urine and the juice he had wrung from a scorpion.

The limit

Valencia's remarkable endurance almost defies belief. Most people die within two days without water in such desert conditions. According to Professor Claude Piantadosi of Duke University, in average outdoor temperatures you can expect to survive about 100 hours without water. Valencia's ordeal sets an upper bound for survival without water, all the more incredible because of the extreme conditions.

030 OUTNUMBERED!

Sally breathes a sigh of relief. Thanks to her powerful new vacuum cleaner with HEPA (high-efficiency particulate arresting) filter, she has cleaned every nook and cranny of her house. There is not a speck of dust anywhere. She has a bare minimum of furniture, arranged so that there are no hiding places for bugs of any kind. Her windows and doors are sealed, and every crack in the baseboards or gap in the floorboards has been filled. No critters of any size can get in, and now she has sucked up every single one that might already be inside. She even has hypoallergenic covers on her pillows and bedding, which have all spent a few days in the deep freeze and then been boil-washed to eradicate any dust mites. Finally, every surface has been swabbed with powerful antiseptic.

'This house is officially bug-free,' she announces proudly to her brother, whom she forced to dress in a sterile jumpsuit and hairnet before he was allowed entry. 'There is not a single bug anywhere inside these four walls.' Her brother coughs. 'Um, that's not strictly true,' he points out. 'There are quite a number of them in here right now . . . a few trillion, actually.' 'What, where?' Sally exclaims in horror. Her brother points to his belly and then to hers.

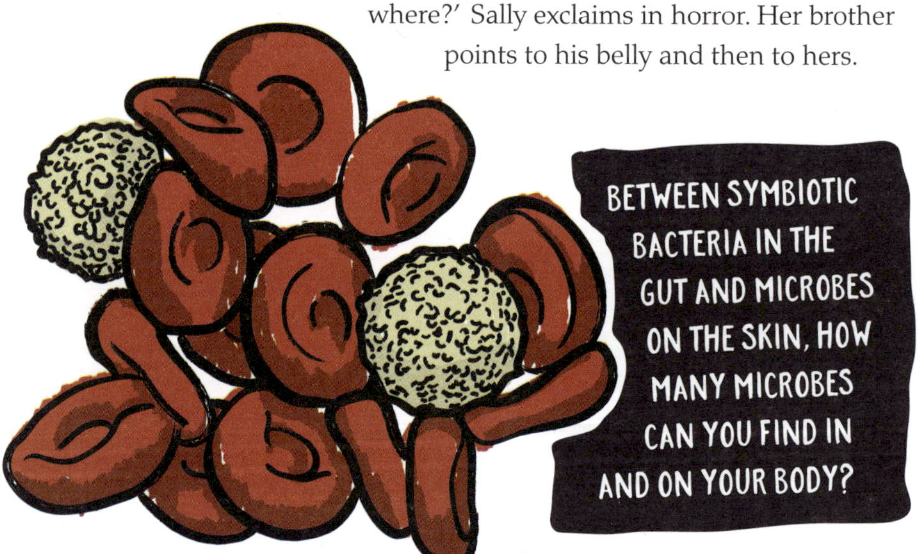

You are what eats for you

The human body is made up of something in the order of 3 x 10^{13} cells – that's 30 trillion. Yet this does not tell even a third of the whole story, because there are three times as many microbes in your gut alone, not to mention billions more on your skin. If you could put them all together, they would be about the size of a football. Without these microbes, you wouldn't last long (your gut flora, for instance, is needed to perform many aspects of digestion that your own cells cannot manage), and the total collection, known as your microbiota, is as unique to you as your fingerprints. This raises interesting questions of identity; the Human Gut Microbiome Initiative suggests that, 'It seems appropriate to consider ourselves as a composite of many species – human [and] bacterial – and our genome as an amalgam of human genes and the genes of our microbial "selves".'

'More bacteria live and work in one linear centimeter of your lower colon than all the humans who have ever lived. That's what's going on in your digestive tract right now. Are we in charge, or are we simply hosts for bacteria? It all depends on your outlook.'

– Neil deGrasse Tyson, *Space Chronicles: Facing the Ultimate Frontier* (2012)

031 WATERED DOWN

Dirk Hogerty has decided to take up a new line of business: selling fishponds to wealthy anglers. Prospective customers are impressed when they see a show pond stuffed with approximately 10,000 fish, but doubts begin to grow as Dirk outlines his working procedures.

'This is not the show pond,' he explains, 'but the only pond'. His intention is to join this pond to a new one next to it, which is to be nine times larger. When the fish have spread evenly throughout the enlarged pond, he will wall it off into ten square ponds, and from each of these he will scoop up all the fish and dump them into one of a series of ten nearby lakes, each of which is 100 times larger than the square pond. After the fish have spread around evenly, these lakes will then be subdivided into 100 squares, and the process repeated. This will result in a series of 1,000 large lakes, each subdivided into 100 squares, giving 100,000 fishponds to be sold to aspiring anglers.

'But the chances of any one of these fishponds, chosen at random, containing even a single fish would be ten to one against,' one of the customers protests. 'The overwhelming probability is that we'd be paying for a pond with no fish in it.' Dirk calmly explains that this doesn't matter: 'The fish will have imbued the lake water with the essence of fishness, and so each pond will be filled with piscatorial energy. In fact, angling in such a pond,' he claims, 'will result in better catches than fishing in a conventional pond that is actually stocked with fish.'

'*Homeopathy may be defined as a specious mode of doing nothing . . . [it] will, at some future time, be classed with historical delusions.*'

– Jacob Bigelow, *Brief Expositions of Rational Medicine* (1858)

Magical thinking

While no one is likely to buy Dirk Hogerty's mystical fishponds, hundreds of millions of people around the world believe in the efficacy of homeopathic remedies and pay out huge sums for them. Few understand the simple mathematics behind the preparation of a remedy, which is based on the counterintuitive and logically absurd principle that diluting an active ingredient makes it more potent.

The dilutions used in popular homeopathic remedies are far greater than in our fishpond analogy. At the '60C' dilution advocated by some practitioners, a patient would have to ingest a volume of remedy equivalent to 10 billion times the volume of the Earth to be sure of getting a single molecule of active ingredient. Homeopaths claim that their remedies achieve therapeutic power by imprinting some sort of ill-defined energy to water, while popular enthusiasm for homeopathy stems from a variety of logical fallacies, errors in critical thinking and psychological bias in assessing evidence.

HOMEOPATHIC REMEDIES INVOLVE PREPARATIONS THAT ARE DILUTED MILLIONS AND BILLIONS OF TIMES, SO IS THERE ANY WAY THEY CAN ACTUALLY WORK?

032 TWO LEGS GOOD

Once the dust has settled and the nuclear winter has passed, the new android overlords of the planet emerge from their fortified factory-bases and set about enjoying the spoils of their decisive victory over the now-extinct humans. They take particular pleasure in nurturing the few surviving remnants of wilderness and wildlife, and helping them to recolonise the Earth.

It is while tending the restored savanna habitat that an android first queries why they continue to manufacture themselves in the image of their biped creators. Other androids linked to the same local network point out the advantages of their bipedal locomotion in the savanna environment: Raised above the tall grass, their visual processing units can more easily scan the environment, while battery management analysis clearly shows bipedal walking to be far more efficient for traversing long distances than a quadrupedal gait.

Androids surveying the nearby forest counter that their anthropoid lower limbs severely hamper their mobility in an arboreal setting, although they do concede that it is useful to have their upper limbs left free to manipulate the environment. The androids agree that the issues of limb design, posture and gait will have to be revisited before the next generation of androids leaves the production line.

MOST OF THE REST OF ANIMAL CREATION GETS ABOUT ON FOUR (OR MORE) LEGS, BUT WHY DO HUMANS WALK AROUND ON TWO?

Ramblin' man

The androids have correctly identified some of the pros and cons of an upright posture and bipedal gait, which paleoanthropologists have identified as perhaps the defining feature of the hominid lineage. Fossil hominids show that bipedalism evolved long before big brains or tool use, which rules out Darwin's suggestion that the need to leave the hands free for tools may have driven the evolution of bipedalism.

The fossils also show that bipedalism evolved in a forest setting, militating against the 'savanna' hypothesis which was also advanced by Darwin and explored by the androids in our story. It is true, however, that for apes anatomically equipped to do it, walking upright is a much more efficient way of getting around: Humans walking on two legs consume only a quarter of the energy that chimpanzees use while 'knuckle-walking' on all fours. Another possible driver for the evolution of bipedalism is that it made for better fighters, since humans can punch harder standing than on all fours, and downward punches are much more forceful than upward ones.

Wading, not drowning

Another intriguing suggestion is that bipedal hominids evolved in woodland next to the coast, and that abundant seafood easily gathered in shallow water was a major contributor to their diet; this is sometimes known as the 'amphibian generalist theory'. If this is true, bipedalism had obvious benefits, as it was the only viable solution for wading.

'[It is] an advantage to man to stand firmly on his feet and to have his arms free.'

Charles Darwin (1809–1892)

033 ASPARAGUSTING

Maisie's dinner party has been going swimmingly, right up until the break between the main course and dessert. A number of the guests excuse themselves to use the restroom, and when they return, two of them are wrinkling their noses and looking very put out.

'Are you both OK?' Maisie inquires. The two guests exchange glances. One hesitantly raises the issue of the smell in the restroom, which prompts an immediate stampede of the remaining guests – all slightly tipsy by this point in the proceedings – to the offending convenience. Of the 16 guests, 15 hold their noses or gasp theatrically, but one looks about her, mystified. She insists that she can't smell anything, and wants to know what all the fuss is about.

The others turn to her in disbelief. 'The odour is unmistakable,' one of them says. 'Surely you cannot fail to notice the overpowering smell of asparagus?' They point out that half of the guests enjoyed the asparagus starter, with the inevitable consequence. At this point the lone hold-out looks even more mystified. 'What inevitable consequence?' she asks.

Meanwhile, the eight guests who have not had the asparagus for starters are pointing accusatory fingers at the other eight, but three of *them* deny responsibility, claiming that they have never in their life been responsible for 'asparagus wee'. This occasions general disbelief all round, until a late-arriving guest, who happens to be an expert in the biochemistry and genetics of olfaction, clears up the whole matter.

'*Asparagus inspires gentle thoughts.*'

– Charles Lamb (1775–1834)

Urine headspace

The lady who cannot smell the asparagus wee is not crazy. Studies suggest that approximately 2 in every 31 people cannot detect the potent odour compounds present in the urine of people who have recently eaten asparagus. Analysis of the 'headspace' (the atmosphere directly above the surface) of the urine of such people shows the presence of volatile compounds such as methanethiol and dimethyl sulphide – sulphur-containing compounds belonging to the thiol family (also found in skunk spray), to which the human nose is extraordinarily sensitive (able to detect concentrations as low as a few parts per billion). These are among the breakdown products of asparagusic acid, a sulphur-containing organic compound found only in asparagus – breakdown products present in the urine headspace after consumption at levels 1,000 times higher than when asparagus has not been consumed. But some studies suggest that around 40 percent of people don't generate these thiols, making it plausible that three of the eight dinner guests who ate the asparagus were indeed innocent of responsibility for the potent wee.

WHY DOES ASPARAGUS MAKE YOUR WEE SMELL?

034 SKIN DEEP

Ra, the ancient Egyptian god of the Sun, is bored. He decides to have some fun, and covers the face of the Sun with glass, which blocks ultraviolet (UV) light. At first the humans do not notice, for the visible light reaching them is not affected, but after a time they began to sicken and die. Children's bones soften and weaken; women's pelvises become deformed and they cannot bear children; the bones of the old become thin and fragile. Only those with lighter skin are spared and thrive, and soon all the population of the land has become light-skinned.

Then Ra, feeling still more cruel and fickle, takes away the glass from the face of the Sun, and the people suffer once more. Their fair skin burns and blisters, their blood is weakened so that they become tired and listless, and those who reach late adulthood are afflicted with malignant growths. Even worse, mothers cannot bear healthy infants, for many perish in the womb, and those that are born have terrible deformities of the spine. This time, only those with darker skin are spared and thrive, and soon all the people of the land are once more dark-skinned.

WHAT MAKES SOME PEOPLE DARK-SKINNED AND OTHERS LIGHT? WHY ARE THERE SUCH PRONOUNCED DIFFERENCES IN SKIN PIGMENTATION AROUND THE GLOBE?

Now Ra reshapes the once-flat Earth into a globe, so that the sunlight reaching the high latitudes will be spread out over a greater area and so become weaker. The people who live near the Equator, where there is little shelter from the fierce tropical sun, remain dark, but those near the poles find that, once again, they suffer from bone diseases. Again, only the lighter-skinned among them thrive, and soon those living near the poles are pale once more. It amuses Ra to play with the skin colour of the humans in this way. But one people, the Inuit, defy him. Travelling into the farthest north, they learn a trick that lets them keep their dark skin, though they barely see the sun for half of each year. And Ra is angered, for he sees the humans have moved beyond his power to control their nature. He grows still more furious when they invent sun cream.

Dark they were

Ra's malicious tinkering mimics the conditions believed to have guided the evolution of skin colour in humans. The great apes are generally light skinned beneath their fur, but when humans became hairless – possibly to boost heat loss through evaporation of sweat, enabling them to keep cool when they moved out into the open grasslands – they were faced with a problem. The powerful UV of the tropical sun leached folic acid from their bodies, risking folate deficiency and related anaemia and birth defects in spinal formation. Therefore they evolved skin rich in melanin, a natural sunscreen.

The paler regions

When humans colonised higher latitudes they faced a different problem: Lack of UV, and corresponding vitamin D deficiency (vitamin D production in the body depends on UV penetration into the skin), leading to rickets and other bone disorders. This selected for lighter skin tones, with less melanin. The Inuit and some other Arctic peoples, however, derive so much vitamin D from their diet of marine animals that they can survive at high latitudes despite their dark skin.

035 INVISIBLE LIGHT

Professor Cavendish has a powerful supercomputer that allows her to test models of animal evolution as accurately as if they had happened in the real world. She decides to run two parallel scenarios through the computer. Each scenario will concern an early hominid (one of our pre-human ancestors), and how well it survives and thrives once a minor tweak has been made to its DNA.

The first scenario that Professor Cavendish runs concerns an early hominid exactly like our real ancestors in every detail, which will serve as a control. The second scenario concerns a hominid that is the same in every respect, except that its eyes are equipped with pigments that respond to infrared and ultraviolet light – light of wavelengths just outside the boundaries of the visible spectrum, beyond the red and blue ends of the spectrum.

Professor Cavendish theorises that the hominid with the extended eyesight will be better equipped to compete in the brutal contest of natural selection, since it will be able to see colours and details that the unenhanced hominid cannot. To her surprise, however, her simulations show that the hominid with the enhanced vision fares less well over time. In evolutionary terms, it proves to be less 'fit'.

SOME CREATURES, SUCH AS INSECTS, CAN SEE ULTRAVIOLET LIGHT, WHILE OTHERS, LIKE SNAKES, CAN DETECT INFRARED LIGHT. WHY DOESN'T HUMAN VISION EXTEND BEYOND THE LIMITS OF THE VISIBLE SPECTRUM?

The economics of evolution

Nothing in nature comes without a cost. Building, running and maintaining body structures and processes takes energy and resources. The organism that makes most efficient use of its resources, getting the best return for the resources it expends, will prosper most. This logic applies to the make-up of the human retina.

The colour of ripe

Humans have evolved vision that is restricted to the visible spectrum for a reason, which is probably that sunlight is far more intense in the visible spectrum than the invisible. Evolving pigments and detection systems for the extremes of the spectrum would be uneconomical and inefficient in evolutionary terms (offering diminishing returns for increasing expenditure of physiological resources). For our hominid ancestors it was probably more important and adaptive to evolve excellent vision for the colours of nutrient-rich ripe fruits, which fall within the visible spectrum. This is why the hominids in Professor Cavendish's second scenario did not prosper: They wasted physiological resources on making pigments to capture light that is hardly there, and which they did not need.

'*Nature does nothing in vain when less will serve; for Nature is pleased with simplicity and affects not the pomp of superfluous causes.*'

– Isaac Newton, *Principia* (1687)

036 WHAT GOES IN . . .

Microman and Captain Tiny – two superheroes who can shrink to microscopic size – decide to have a rather unusual race. They creep onto the plate of an unsuspecting man who is about to eat a roast chicken with vegetables, and each choose a piece of food in which to hide. Microman makes straight for the Brussels sprouts, while Captain Tiny heads for the potatoes. Both are swallowed at the same time, and as they hurtle down the diner's esophagus towards his stomach, Captain Tiny calls out to her friend, 'See you in the toilet bowl. I'll be waiting for you!' 'Not if I get there first,' Microman replies, splashing into a soup of gastric juice and macerated food.

At first Microman curses his choice of carrier food: The tough, fibrous sprouts resist dissolving in the acid and float about, mostly intact. Meanwhile, the soft, starchy potato quickly dissolves, and a bolus of chyme (partly digested food) passes through the pyloric sphincter and into the duodenum, carrying Captain Tiny with it. When Microman finally catches up with her in the colon, having taken a considerable time to get through the small intestine, she looks fed up, having been stuck in a fold for hours. From here the higher fibre content of the sprouts begins to pay off, as Microman finds himself packed into a pliant but coherent sausage of fecal matter that moves through the colon at a relatively fast pace. There is a bit of a holdup in the rectum, but Microman catches the unmistakable scent of volatile aromatics from coffee, followed by a headlong rush into the toilet. Able to get clear of his vessel, Microman hangs about under the rim of the toilet until his friend arrives later the same day. 'You win,' admits Captain Tiny, ruefully. 'In fact, I think you made it in record time.'

'*Digestion, of all the bodily functions, is the one which exercises the greatest influence on the mental state of an individual.*'

– Jean-Anthelme Brillat-Savarin (1755–1826)

Time in motion

Different foods travel through the digestive system at different rates in different people, but the average gastrointestinal transit time, as it is known, is a surprisingly long 40 to 50 hours. It takes just a few hours for food to move from the stomach into the small intestines (the duodenum, jejunum and ileum), where the bulk of digestion occurs over around 3 to 10 hours. The majority of the time is spent transiting the bowel or large intestine, and chiefly the colon; here bacteria digest food components that our own cells cannot manage, water is reabsorbed and feces are formed. This usually takes 30 to 40 hours, but can take much longer in people with constipation, poor diet or poor bowel health. You can get an idea of your own transit time by eating a good few spoonfuls of something that will pass through your gut relatively unharmed and be readily identifiable at the other end, such as sweetcorn or sesame seeds.

EXACTLY WHICH MEAL IS IT THAT YOU ARE POOPING OUT WHEN YOU GO TO THE TOILET? THAT MORNING'S BREAKFAST, YESTERDAY'S LUNCH? THE ANSWER MAY SURPRISE YOU.

037 THE COLOUR OF RUNNY

Jane and Laura are having a dispute. 'Is not,' says Jane, stamping her foot. 'Is so,' replies Laura, stamping hers. 'Now, now, children, what are you fighting about?' asks Dr Goodman. 'Jane says that her bogies are better than mine because they are all yellow,' pouts Laura. 'Well, Laura says her bogies are the best because they are bright green!' 'Easy now, girls, there's no need to fall out over the colour of your nasal mucus,' chortles Dr Goodman, 'but I do have to tell you that neither of those colours is particularly salubrious' 'Salubriwhat?' puzzle the girls.

Dr Goodman leads them into her office and points to the chart on the wall that shows a section through a person's nose and sinuses. 'This is where you breathe in air but, to wet the air so it doesn't dry out your lungs, the lining of your nose and sinuses makes a sticky, gloopy fluid called mucus (what you call bogies or snot). The mucus is mainly water, with some colourless stuff in it, like salt and protein. When you are healthy it is nice and clear. But when nasty germs like bacteria take up residence in your nasal passages, your body sends special defensive white blood cells to fight them.'

The girls exchange looks: 'But why are our bogies all colourful?' 'Having a whole load of white blood cells in your mucus makes it yellow, so I'm afraid, Jane, that you may be just at the start of a bacterial infection. And Laura, you may be nearer the end of one, because the green colour in your mucus shows that the white blood cells have been using their secret weapon, an enzyme called myeloperoxidase, which causes the green colour.'

The girls consider this for a while, and then Jane speaks up: 'So what you're saying, doc, is that Laura has given me a cold?'

'Did not!'

'Did too!'

NASAL MUCUS – A.K.A. SNOT – CAN BE CLEAR, WHITE, YELLOW OR BROWN, BUT THE COLOUR WE MOST ASSOCIATE WITH BOGIES IS GREEN. WHY SHOULD NASAL MUCUS ADOPT THIS UNUSUAL COLOUR?

Tracking down the green stuff

Myeloperoxidase (MPO) is a destructive enzyme capable of killing cells, and is released by white blood cells called neutrophils as they do battle with invading microbes. It was first identified in 1941 by Kjell Agner (1913–70), a Swedish physician who wanted to identify the nature of the green substance in pus and phlegm, particularly the mucus produced in the lungs of tuberculosis patients.

Heme colour

MPO contains a heme group – an organic compound containing an iron atom. Iron, in its various states of oxidation, can produce a range of colours, including the red of hemoglobin and, in MPO, green. For some reason this green colour only becomes apparent when the MPO is released by expiring neutrophils into the surrounding environment (in this case, mucus), so green snot generally indicates a battle with infection that is well underway.

038 BRIGHT LIGHTS, BIG SNEEZY

'I've gathered you all here, in this room, in order to reveal two important facts,' Inspector Dubois tells the assembled group, which includes René, the Comte's mechanic; Emile, the lighthouse keeper and childhood friend of the Comte; Madame Choufleur, the Comte's former governess; and the Comtesse herself. 'Firstly, that the Comte was murdered, and secondly, that I have identified the murderer!' There are gasps all around; only the Comtesse seems unmoved, remarking icily, 'My late husband died in a traffic accident, Inspector.' 'Indeed he did, Madame,' the Inspector admits, 'except that I do not believe it to have been an accident at all. Allow me to explain how I reached this conclusion.'

The first clue was the testimony of the Comtesse herself, who remarked that moments before she heard the crash she had been dazzled by a bright light apparently shining from the top of the nearby lighthouse. The second clue came from René, who insisted that the Comte was an extremely cautious driver, and not at all accident-prone, and the third from Madame Choufleur, who recalled that her pet name for the Comte when he was a little boy had been 'Sneezy Sam', just like his father before him.

'And so I have concluded that the murder was arranged with diabolical genius by one who knew of the Comte's peculiar affliction: Autosomal-Dominant Compelling Helio-Ophthalmic Outburst, or ACHOO. This condition sets off sneezing when the sufferer experiences a sudden change in the intensity of light – sneezing violent enough to cause a car crash. Normally sunlight is the trigger, but any bright light will do – a bright light such as the one a lighthouse keeper is able to direct.' Emile snarls and leaps to his feet, but before he can flee, Inspector Dubois whips out a flashlight and shines it right in his eyes. The murder suspect helplessly sneezes three times in a row, delaying him long enough for handcuffs to be applied. 'It seems that the condition is more common than one might think,' observes the Comtesse.

Photic sneezing

More commonly known as photic sneezing, this inherited reflex response to sudden changes in the intensity of light is believed to affect 17 to 35 percent of the population. First noted by Aristotle – who asked in his *Book of Problems* (c. 300 BCE), 'Why does the heat of the Sun provoke sneezing, and not the heat of the fire?' – photic sneezing remains unexplained. Possibly it is caused by crossed wires in the trigeminal nerve, which innervates the eyes, nasal cavity and jaw, such that strong stimulation of the nerves controlling the pupil (which contract sharply in response to sudden brightening) accidentally mimics strong stimulation of the nasal cavity, triggering the sneeze reflex. An alternative but related explanation is 'parasympathetic generalisation': Both pupil contraction and sneezing are controlled by the parasympathetic nervous system, so stimulating one may trigger the other.

039 HAMMER TIME

Big Joe is eating osso bucco in the back room at Mazelli's when Little Mikey Ninefingers bursts in. 'Whaddya got for me, Mikey?' he growls. 'Oh boy, Joe, you ain't gonna believe what I got for you. You remember our friend, who had the trouble with the contractors?' 'How'm I gonna forget when he's stinking up my basement?' grumbles Paulie Shoeshine. 'Well, I just came up with an idea for how to get ridda' him, and not just him, but anyone we, uh, have a problem with in future. I just saw this freaky website, and it gave me a flash of inspiration!' 'Get to the point,' growls Big Joe.

'So there's this company, see, and they say they can bury people like you never seen before. What they do is, they freeze the body solid with liquid nitrogen, and then they shatter it into, like, a million tiny pieces, and they turn them into freakin' compost, and grow trees in 'em. So I says to myself, why can't we do the same? Liquid nitrogen is real cheap, see, and all we have to do is

IF YOU FROZE YOUR HAND IN LIQUID NITROGEN AND HIT IT WITH A HAMMER, WOULD IT SHATTER?

freeze our friend in Paulie's basement until he's like a big piece of glass, and then we shatter him into tiny pieces and spread him around down at Old Piscatelli's farm. Completely untraceable.'

The mobsters look at Big Joe. He puts a piece of veal in his mouth and chews slowly. 'Don't sound feasible. I seen them on TV, freezing a flower and shattering it, but a human body ain't no flower. We gonna have to test it,' he growls. 'Little Mikey, go down to the yard and pick up some o' this liquid nitrogen. And get a hammer, a big one.' After Little Mikey Ninefingers has gone, Paulie asks, 'So how we gonna test it, Big Joe?' 'We're gonna see how easy it is to make him Little Mikey Fivefingers.'

Going brittle

Pretty much every material known will cross what is known as the ductile–brittle transition threshold, if it gets cold enough. As materials cool, there is a decrease in the ability of their constituent atoms to shift and transfer deformation stress to neighbouring areas, so that energy is concentrated in one region and bonds between atoms simply break. Liquid nitrogen has a temperature of almost –200°C (–392°F), so it will indeed freeze to extreme brittleness almost anything with which it comes into direct contact, but this is only half the story.

Body of evidence

A human hand, let alone a whole body, is a thick, dense structure. It would take some considerable time for the freezing effect of exposure to liquid nitrogen to penetrate right through. Even then, the sheer mass of the frozen hand would require a great deal of energy to break, and it would likely only break at the actual impact site. Shattering it into millions of pieces, as seen in films, is implausible. It has been suggested that shattering might be achieved by combining freezing with some other process, possibly involving ultrasound, but this has yet to be convincingly demonstrated.

040 BLIND SIGHT

The nurse was about to peel off the bandages from the boy's face when the surgeon asked him to wait for a moment, so that she could attempt an experiment. She asked the boy to handle two small wooden objects, which she named as a cube and a sphere, making sure he knew which was which. When he had thoroughly familiarised himself with their tactile properties, she took them from him and placed them on a table a short distance away. Then she bade the nurse continue with removing the bandages.

The child had recently been operated upon to remove cataracts that had blinded him since birth; this moment would be the first time he would perceive something visually. Allowing him a time to grow used to the unaccustomed brightness, the surgeon now asked the boy to direct his newly restored vision to the objects on the table. Could he distinguish one from the other?

Molyneux's problem

This scenario replicates a thought experiment proposed by the Irish philosopher William Molyneux (1656–98), in a 1688 letter to the English philosopher John Locke (1632–1704), which became known as Molyneux's problem. It came to be one of the most contested questions in 18th-century philosophy. Molyneux and Locke, empiricists who believed that knowledge and ideas can arise only through experience, which in turn depends on sensation, averred that the boy would not be able to tell one from the other with sight alone. Experiments exactly like the one described, performed in 1728 by surgeon William Cheselden (1688–1752), and again on a series of children in India in the 21st century, seem to support their claim, since the children could not immediately distinguish the objects, although they quickly learned to do so.

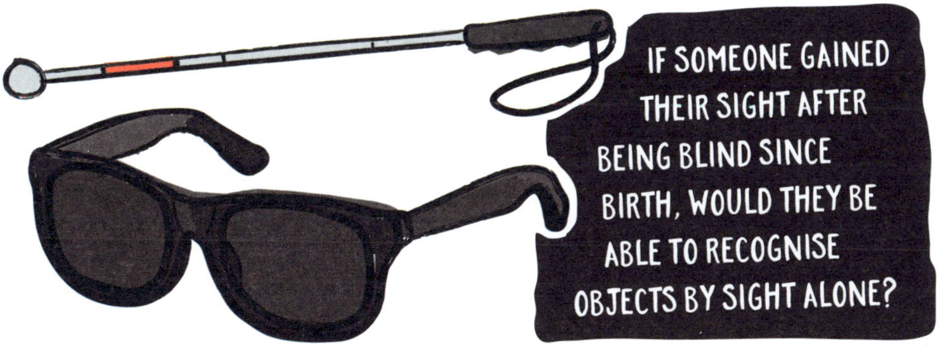

Colour blind?

A related question is whether someone blind from birth can have any true concept of colour. Empiricists believe not, since they say the only way to learn such a concept would be through sensing it. Philosophers of consciousness argue over whether it would be possible to acquire such a concept through somehow learning all the physical facts about colour (including the minutiae of neurological correlates of such perception), or whether subjective experience is necessary, which implies a form of dualism (the belief that there is a distinction between the worlds of the purely physical and the mental).

'Suppose [a] blind man made to see [a] cube and [a] sphere placed on a table . . . query: Whether by his sight, before he touched them, he could now distinguish and tell which is the globe, which the cube?'

– William Molyneux, *An Essay Concerning Human Understanding* (1694)

041 CAN A COMPUTER THINK?

Dorothy, the Scarecrow, the Tin Man and the Cowardly Lion approached the giant curtain. 'ASK YOUR QUESTIONS, AND I, THE GREAT AND TERRIBLE OZ, WILL ANSWER THEM!' a great voice boomed. The four travellers huddled together in terror, but Dorothy bravely spoke up, 'If you please, Oz, the Great and Terrible, can you tell me how to get home?' There was a pause, during which could be heard the grinding and clanking of gears, and then the voice boomed out once more: 'COMPUTING ERROR AT LINE 5; REPHRASE THE QUESTION.'

Toto, Dorothy's little dog, ran forward and began barking at the fringe of the curtain, then seized it in his jaws and tugged. The whole curtain slid to the ground, revealing a huge apparatus of gears, levers, belts, cogs and pins. 'Why, it's nothing but a machine,' said Dorothy, wonderingly.

The Tin Man pointed to a long sheet of paper covered with tiny holes, which whirred through a series of rollers before tripping a series of tiny levers, like the workings of a grand piano. The Scarecrow scratched his head: 'But how can it know what you mean by "home"? How can it know what any of us mean by anything?' The voice sounded again: 'YOUR INPUTS ARE PROCESSED INTO OUTPUTS.'

'I may not have a brain,' protested the Scarecrow, 'but it seems evident that your processing is purely syntactical, without any semantic value. How do we know if you're really a thinking being at all?' 'I SOUND CONVINCING, AND IF YOU CAN'T TELL THE DIFFERENCE, PERHAPS THERE ISN'T ONE.'

The Room, the Game and the Mill

Dorothy and friends have just encountered an amalgam of several different thought experiments that philosophers have used to explore the question of machine or artificial intelligence, and by extension of the nature and knowability of consciousness itself. In John Searle's (b. 1932) Chinese Room thought experiment, a man in a sealed room answers questions in Chinese, a language he cannot speak or understand, by processing inputs into outputs using a manual. Can he, or the Room itself, be said to understand Chinese?

In Alan Turing's (1912–54) Imitation Game test, a computer answers questions via a text interface; if human interlocutors cannot tell that it is not a person, can it not be said to be intelligent?

In Gottfried Wilhelm von Leibniz's (1646–1716) Mill thought experiment, a thinking machine is blown up to the size of a mill or factory, and inquirers are invited to search inside it for anything that can explain thought or consciousness.

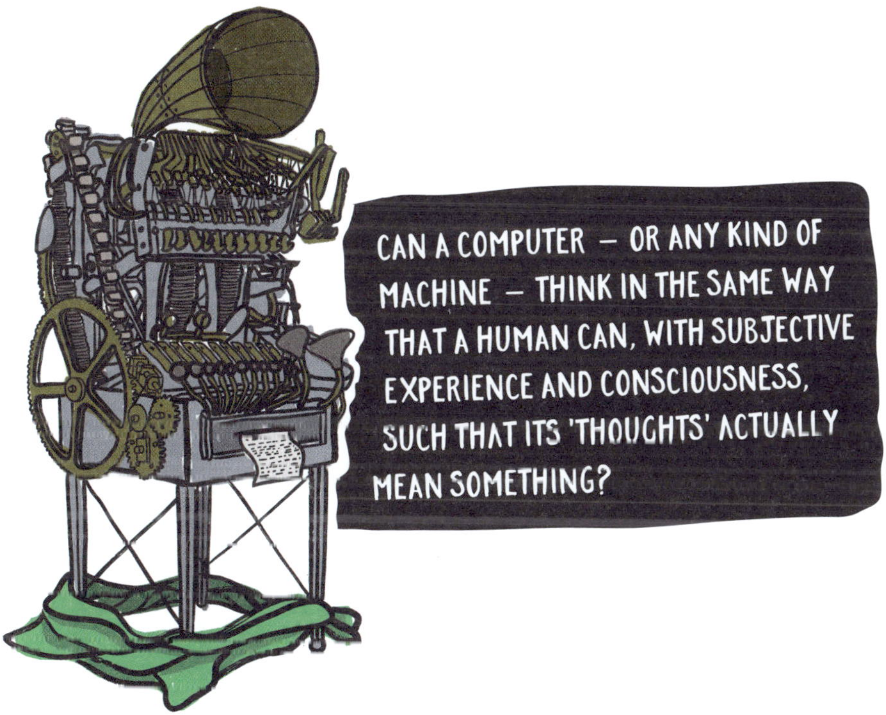

042 PET SOUNDS

Mittens was sceptical. She insisted it was just a coincidence. 'If you go to the door enough times,' she purred, 'then you're bound to be right occasionally. Even a stopped clock tells the right time twice a day.' Patches growled at her: 'You're just jealous because you can't tell when Mary's coming home.' Mittens licked a paw demurely: 'So you mean to tell me that whenever I see you leaping about like an overexcited footstool, it's because your "psychic sixth sense" has gone into overdrive?' 'Exactly,' he barked, wagging his tail. 'I can tell when she's on her way home. And I can prove it.'

Patches told Mittens to keep a careful record of the time of day that he started to jump about by the front door, and the time of the day their owner, Mary, arrived home. After a week the cat had to admit that there appeared to be a correlation between the two, since on six out of the seven days Patches' excitement had preceded Mary's arrival by about 6 minutes. 'This proves nothing,' she meowed. 'If Mary gets home at the same time every day, even a goldfish could predict when to sit by the front door.' Patches pointed out that Mary was a shift worker, and came home at unpredictable times of the day and night. Mittens flicked her tail: 'Hmm, I'm not convinced. On Thursday you were caught by surprise, and that was the day it was raining so heavily. I think you can hear or smell Mary from a long way off, and that's what tipping you off. The one day the weather interfered with your little trick, it didn't work.'

'Many owners of dogs, cats, horses, parrots and other animals find their animals pick up their thoughts and intentions.'

– Rupert Sheldrake, *Dogs That Know When Their Owners Are Coming Home* (2011)

Morphic malarkey?

Parapsychologist Rupert Sheldrake (b. 1942) believes that animals are attuned to a quasi-mystical 'morphic field' that permeates nature, and which transmits information in ways we label 'psychic'. He suggests pet owners prove it by trying this experiment for themselves. Sceptics point out that it is very hard to perform such an experiment in a properly controlled, double-blind fashion that can rule out confounding factors such as those identified by Mittens, and they dispute that trials carried out in this fashion provide any evidence for Sheldrake's claims.

DO DOGS KNOW WHEN THEIR OWNERS ARE COMING HOME? ALTERNATIVE RESEARCHER RUPERT SHELDRAKE HAS SUGGESTED THAT THE BEST PROOF FOR THE EXISTENCE OF PSYCHIC POWERS MIGHT BE TO SHOW THAT THEY DO

043 TEENAGE MELTDOWN

Sigmund slammed the door, then opened it to shout, 'I hate you!' before slamming it again. 'But, Siggy,' pleaded his mother, 'we're only thinking of you.' 'You don't understand me,' he shouted through the door, 'and don't call me "Siggy": I'm not a child any more!'

Sigmund flung himself down on his bed and stared at his posters of Darwin and Helmholz. Why couldn't his parents understand that things couldn't go on as they had done before? He brooded on his conflicted feelings. On the one hand he wanted his parents to share his excitement about the new scientific discoveries, and to support his ambition to become a doctor. On the other hand he despised their middle-class sensibilities and antiquated morality, and wanted nothing more than to get away, to be his own person. His parents were so cloying, they suffocated him. They didn't really care about him, only themselves! If only he could find a mentor, someone he could look up to and follow.

WHY ARE TEENAGERS SO GRUMPY? IS IT INEVITABLE THAT TEENAGERS WILL BE GRUMPY AND MOODY, AND WHAT ARE THE REASONS FOR THIS SUPPOSEDLY TYPICAL BEHAVIOUR?

The storm years

Poor Sigmund displays all the signs of what is known as the 'classical theory of adolescence', which views the teenage years as a time of *Sturm und Drang* (storm and stress). According to Erik Erikson (1902–94), one of the leading figures of the psychoanalytic movement after Freud, adolescence is characterised by identity confusion, which leads to identity crisis. According to child psychoanalyst Peter Blos (1904–97), the adolescent 'disengages' as he or she tries to forge an independent sense of identity. Disengagement produces regression, as when adolescents look for substitute parents through hero worship (e.g. of sports or rock stars), and ambivalence, in which the teenager simultaneously needs and rejects parental love and approval.

The kids are alright

The problem with the classical theory is that it is not supported by the evidence. Surveys show that the majority of teenagers are well adjusted, rather than afflicted by raging psychic conflict, and neither they nor their parents report major conflict between the generations.

'At no other phase of the life cycle are the pressures of finding oneself and the threat of losing oneself so closely allied [as during adolescence].'

– Erik Erikson (1902–94)

044 OPEN WIDE!

Agent Arek scanned the detention room with four of her sixteen optic pods. It held at least 200 humans, and to her they all looked the same. Working out which one was actually a rogue android could take some time. Agent Bismet entered the control booth and undulated his tentacles quizzically: 'Any developments?' Arek vibrated her optic pods in noncommittal fashion. 'The only thing I've noticed is that a bunch of them keep performing this strange behaviour – look, there's one doing it now.' One of the humans opened his mouth to its full extent, stretching his jaw, closing his eyes and inhaling deeply. His neighbour promptly followed suit.

'Oh, I know what that is.' Bismet flushed purple. 'I've seen it before on Earth – they do it all the time. It's called yawning.' 'Perhaps it helps them draw more oxygen into their lungs?' suggested Arek. Bismet bobbed his optic pods negatively: 'If that were true, wouldn't they do it after strenuous exercise, not when sitting around? And besides, I read that human fetuses do it in the womb. No, I think it must be to do with being bored or sleepy – you said they'd been sitting in there for hours.' Arek used the sensors to scan for brainstem arousal markers; it was true that many of the humans' brainstems were signalling low arousal levels. She noticed also that the yawning individuals displayed slightly elevated brain temperature.

'What's really interesting is that the behaviour is clearly contagious. See how many individuals repeat the yawn when they notice someone else do it. It must be some empathic response linked to their theory of mind – their ability to empathise with others.' Arek slapped her tentacles triumphantly. 'That's it! Androids look exactly like humans, down to the molecular level, but they're essentially zombies, without any theory of mind or consciousness of any kind.' She activated one of the screens in the detention room to display a video of a human yawning extravagantly. Within minutes almost every human present had followed suit – all except one. Arek activated the disintegrator and atomised the suspect. 'Er, was that not a little hasty?' queried Bismet with a coil of his tentacle. 'Perhaps he simply wasn't sleepy?'

Ubiquitous yawning

Yawning is contagious in primates, although studies suggest that only around half of adult humans reliably display this effect, so Arek was indeed acting prematurely. Contagion seems to be linked to theory of mind, the psychological mechanism that allows one mind to 'put itself in the shoes' of another. Autistic people and under-fives are said not to yawn contagiously. Yawning is not contagious in many other species, but the behaviour itself seems to be almost ubiquitous in the animal kingdom; even birds and fish gape in a yawn-like fashion.

Theories of yawning

The traditional theory that yawning is about drawing in extra oxygen has been debunked, since yawning has been shown not to affect levels of oxygen in the bloodstream. It is now thought to be linked to changes in arousal, as when people are bored or nervous. A more recent theory is that yawning is a reflex response to overheating in the brain, intended to draw in cool air to help with thermoregulation.

045 TIES THAT BIND

Annabel, Chimi and Duri surveyed the toys laid out on the blanket. Chimi was the first to leave her mother's side and crawl over to the playthings. Eventually Annabel and Duri joined her, inquisitively turning over the blocks and dolls and sticking most of them in their mouths. A strange woman came into the room and sat down to play with them. Chimi eyed her suspiciously. Duri chewed on a plastic brick. Annabel glanced anxiously at her mum.

Abruptly, all three of the mums got up and left the room, leaving the three babies alone with the stranger. Annabel immediately started to cry. Chimi was very anxious. Duri barely batted an eyelid. Then the stranger left too, so that the little ones were on their own. Now Chimi also began to cry. Finally all three mothers returned. Annabel and Chimi crawled over to theirs as fast as they could, but when Annabel's mum picked her up she started to struggle and push her away. Duri continued to focus on the toys, and ignored his mother's calls.

Suddenly an evil scientist burst into the nursery, snatched the three children, and whisked them off to his laboratory. He placed each one in a room with, on one side, a bare wire frame that dispensed milk and sweet treats, and on the other, a simple loop of wire wrapped in towelling, which dispensed nothing. After getting over their initial shock, the babies enjoyed the sweet treats from the wire frame, but they kept their distance from it and stayed close to the towelling loop. When the fiendish scientist hit them with a blast of compressed air, the startled infants clung to the towelling loop. At this point the police broke down the door, rescued the babies and carted the mad scientist off to jail.

Strange situation

Annabel, Chimi and Duri have been subjected to two landmark experiments in the history of psychological research on infant–caregiver attachment, each controversial in its own way. In the 'strange situation' experiment, babies were classified according to their reaction on being left alone with a stranger. This led to the controversial categorisation of children by the style in which they expressed attachment to their caregiver, with some infants labelled 'anxious–avoidant' and others 'ambivalent'. In a 1959 study infamous for its cruelty, Harry Harlow (1905–81) raised infant monkeys with crude surrogate mothers that were either bare wire frames dispensing milk or crude towelling dolls that gave nothing. The baby monkeys preferred the latter, especially when stressed, apparently valuing even the most pitiful degree of tactile comforting over resource provision, a finding that struck at the prevailing orthodoxy that infant attachment behaviour – such as crying when mother leaves the room – is little more than conditioned cupboard love.

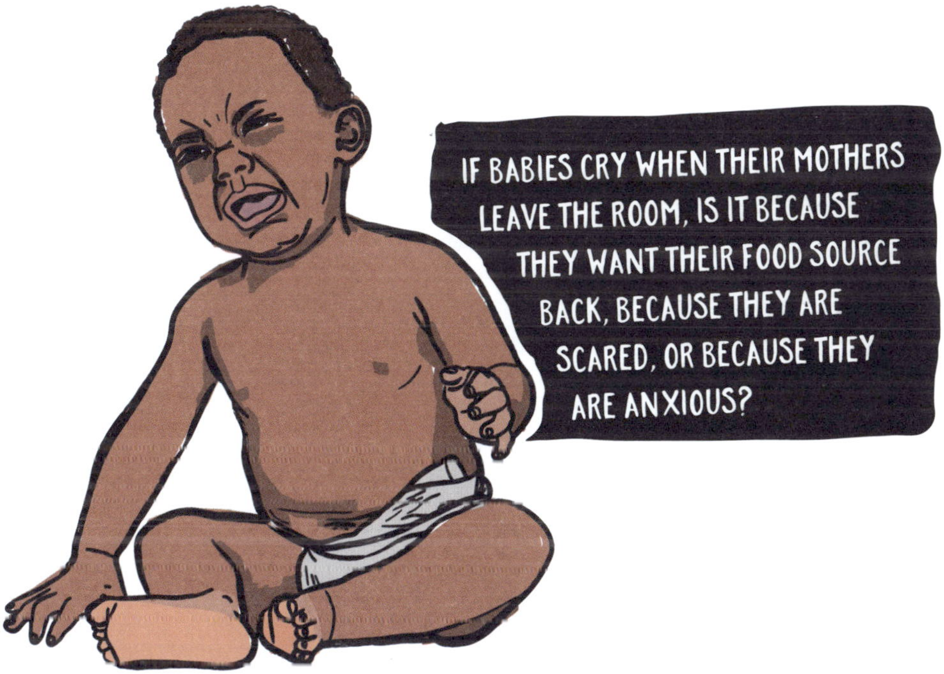

046 THE COCKTAIL PARTY EFFECT

Wanda was not enjoying the party at all. Sure, the guests were glamorous, and some of the frocks were to die for. The drinks were being mixed by the most divine bartender, and the setting itself was everything you could wish for. The nibbles were heavenly, the music loud and exciting and the hosts kind and hospitable. All her best friends were there, everyone she didn't know looked fascinating, and Freddie, with whom she was in close conversation, was saying the most interesting things. But she was having a terrible time.

The problem was that every time she caught the thread of Freddie's conversation, she would get distracted and lose it again, and all because people absolutely, positively would not stop saying her name – at least, that's what it sounded like. First it had been the couple over by the patio doors: '. . . I know, I wonder that too sometimes . . .' Then the guy with the afro by the piano: '. . . and then I thought, I'll just wander back that way . . .' Next it had been Dr Margolies who was at the bar: '. . . Well, it's no wonder they feel that way . . .' And then it was the very tanned lady in the kitchenette: '. . . It's the eighth wonder of the world, I tell you . . .' Wanda's head was spinning from turning this way and that, and it wasn't until Freddie said her name quite loudly that she realised he had asked her something three times in a row. 'I'm so sorry, Freddie darling, you know I never could abide cocktail parties'.

'Cocktail party: A gathering held to enable forty people to talk about themselves at the same time. The man who remains after the liquor is gone is the host.'

– Fred Allen (1894–1956)

Unconscious processing

First described by acoustic engineer Colin Cherry in 1953, the cocktail party effect is striking evidence that there are various levels of conscious attention and processing. Wanda cannot possibly pay attention to every conversation in the party, yet at some level she must be doing exactly that, since when her name – or something that sounds like it – crops up, she is immediately able to 'tune in' to the relevant conversation. She is even able to be aware that Freddie has repeated the same question thrice, although she has no idea what he has actually asked.

Blindsight

An even more striking demonstration of variation in the conscious availability of perceptual processing is the phenomenon of blindsight. This is where a person appears to have gone blind for purely psychological reasons, and although vision is not available to their conscious awareness, they can point to something or change their gait to avoid an obstacle.

THE RULES OF THE UNIVERSE

'The only way of discovering the limits of the possible is to venture a little way past them into the impossible.'

– Arthur C. Clarke (1917–2008)

047 WEIGHTY MATTERS

Crafty Coyote was determined to catch that pesky ground-cuckoo, and he was sure that this time the infernal bird would not get away. His elaborately laid plan was flawless, and now the delivery man from Zenith had delivered the final piece of the jigsaw: a 1 metric ton (1.1 ton) weight complete with massive chain. A loud honk signalled the imminent arrival of the ground-cuckoo, and Crafty hid behind a large rock at the edge of the cliff.

Exactly as planned, the speeding bird shot straight through the large piece of paper cunningly painted with a receding desert motorway, and plunged over the cliff. At the same moment, Crafty heaved the 1-ton weight over the cliff as well, cackling as he imagined the crushing blow it would deliver when it landed on top of the bird. Unfortunately he overbalanced and found himself plummeting through the air alongside both weight and bird. Determined not to panic, Crafty tried to think through the problem logically.

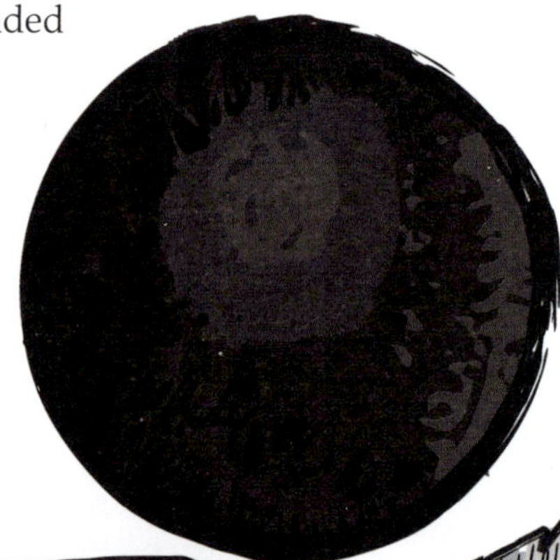

A CANNONBALL IS CLEARLY HEAVIER AND MORE DENSE THAN A FEATHER, SO THAT A VAST DISPARITY IN DOWNWARD FORCE IS SENSED WHEN HOLDING ONE IN EACH HAND; SO WHY DOES THE CANNONBALL NOT FALL FASTER (IN A VACUUM) THAN THE FEATHER?

First he thought of shackling the weight to the bird's ankle, to weigh him down so that he would fall faster and Crafty could land on top of him. Then he realised his mistake: The lighter bird would simply slow down the heavy weight, and Crafty would land before either of them. But then, combining bird and weight cannot speed up their fall at the same time as slowing them down. He was still pondering the paradox when the ground-cuckoo opened a parachute and floated serenely down to Earth; Crafty just had time to notice that he and the heavy weight were about to land at the same time, in exactly the same place . . .

Galileo's experiment

Crafty Coyote has just explored the fundamental paradox at the heart of the Aristotelian physics of falling objects. Following Aristotle (c. 384–322 BCE), the received wisdom up until the 17th century was that heavier objects fall faster. This seemed to accord with common sense, in that the evidence of the senses showed that a heavy object exerts more downward force. But as Galileo (1564–1642) pointed out in a thought experiment, this logic is self-contradictory, for it follows that combining a heavy object with a lighter one will both slow it down – because the lighter one falls more slowly and will thus hold back the heavy one – and speed it up, because the combination of the two objects has a greater overall mass. He pointed out that the only solution is that all objects fall at the same rate, regardless of mass, and he proved it by dropping balls of different masses from the Leaning Tower of Pisa.

Moon drop

In the Earth's atmosphere, air resistance dramatically affects the rate at which things fall, so that a feather and a cannonball will not actually fall at the same rate. But in a vacuum, Galileo's principle of equivalence can be dramatically demonstrated, as when astronaut David Scott (b. 1932), during the Apollo 15 mission to the Moon in 1971, dropped a hammer and a feather to show they fell at the same rate.

048 NO SPEEDING

Cosmicman has god-like powers of strength and speed. He can fly faster than a speeding rocket, and leap over mountains with a single bound. Bored with crushing petty criminals on Earth, and poorly educated in physics, Cosmicman decides to take on the ultimate challenge and fly faster than the speed of light. Blasting into space, he accelerates towards the edge of the Solar System. Before he has even passed the Moon, he is travelling faster than any human-made object has ever gone. By the time he reaches the orbit of Mars he is travelling at 10 percent of the speed of light. Martians observing him through powerful telescopes notice that the watch on his wrist seems to be running noticeably slower, due to time dilation, as predicted by Einstein's relativity theories. To Cosmicman, on the other hand, looking at Mars with his cosmicvision, it is the Martians who are moving in slow motion, and their clocks that are running slow.

By the time Cosmicman passes Pluto he is travelling at 86.5 percent of the speed of light and the universe looks increasingly strange to him. Everything else has slowed dramatically, and what's more everything else seems to be shrinking, or at least contracting along the same axis in which he is moving. He hurtles past a stationary Plutonian rocket ship, which he knows to be 200m (656 feet) long, but to him it looks to be just half as long (although its diameter has not changed). To the Plutonian rocketeers, Cosmicman himself appears to be half as tall as normal.

As he reaches the heliopause – the boundary between the Solar System and interstellar space – Cosmicman has achieved 99 percent of the speed of light but he is finding it increasingly hard going, since he appears to be getting heavier and heavier, thanks to relativistic mass increase. Making a supreme final effort, he accelerates to 99.99 percent of the speed of light – by which time he appears to an interstellar alien bystander to be only around 13.5cm (5.5 inches) tall – but the more energy he exerts the heavier he gets, and so the more energy it takes to accelerate further. Exhausted, Cosmicman gives up, stops accelerating and coasts back to Earth (which only takes him around 14 hours). Only as he barrels past Earth towards the Sun does he remember that he now has to expend the same amount of energy to decelerate to a standstill.

The ultimate limit

Cosmicman abandoned his hubristic quest because the truth finally dawned on him: If he did reach the speed of light, time itself would come to a standstill, and if time freezes, so does movement. At the same time, his mass would have become infinite, and in order to reach this point he would have to have expended infinite energy. He vows to invent a cunning device that will transform him into a massless particle like a photon, enabling him at last to reach the ultimate speed limit.

049 THE BOOK THAT NEVER WAS

Billionaire Nazi fetishist Adnan Forbes celebrates his successful bid for the recently uncovered first draft of *Mein Kampf* by hatching an audacious plan. He will use his vast wealth to build a time machine that will transport him back to the time at which it was written, in 1924; then he will sneak into Landsberg Prison and get the future dictator (and then prisoner) Adolf Hitler to sign it for him. As Forbes sets the dials on his time machine, a young radical named Cody sneaks into the transporter chamber and hides herself. Billionaire and temporal stowaway both travel back to 1924.

Forbes uses his extensive historical knowledge to bribe his way into the prison, but he is stunned and disturbed to discover that the young Hitler has never heard of *Mein Kampf* and has never written anything remotely like it, although the imprisoned Nazi is intrigued to begin reading the manuscript the time traveller has brought. Forbes is still pondering the meaning of this development when Cody pops up behind Hitler with a loaded gun, and puts it to the would-be demagogue's head.

'What are you doing?!' cries Forbes. 'This evil monster is my great-grandfather,' declares Cody. She cocks the gun and prepares to fire: 'Now I will strike the blow that redeems the shame of my family and saves the world from years of misery and horror.'

SOME INTERPRETATIONS OF PHYSICS RAISE THE INTRIGUING POSSIBILITY THAT IT MIGHT BE POSSIBLE TO TRAVEL BACK IN TIME, BUT WHAT PARADOXES MIGHT RESULT FROM SUCH A JOURNEY?

'Stop!' implores Forbes. 'Don't you realise? If you kill him before he has sired your grandparent, then you will never have existed, in which case how could you have travelled back in time to kill him? The resulting paradox could destroy reality.'

An unusually well-informed guard bursts in, warning, 'Your mere presence is already causing a calamitous paradox: You know that Hitler was not assassinated in 1924, so it should be impossible for you to kill him, yet it would self-evidently be possible for you to pull the trigger and finish him off now.' All this time Hitler has been avidly reading the pages of *Mein Kampf*, and now he leaps to his feet, crying, 'Mein Gott! This is genius. Who wrote it?'

Grandfather and other paradoxes

Forbes and Cody have set in motion a combination of several paradoxes that could be triggered by travelling back in time. Cody encounters the classic 'grandfather paradox', in which a time-travelling assassin would violate causality so that his or her own existence is prevented, which in turn would prevent the violation of causality. She also runs up against a related paradox, in which it seems simultaneously to be true that she can and cannot assassinate Hitler. Meanwhile, Forbes has engendered what is known as an ontological paradox, in which causality is violated by creating a closed loop of time: The manuscript of *Mein Kampf* exists in the future only because he took it back to the past. History will attribute authorship of the book to Hitler because it has fallen into his hands and he will claim it as his own, but where did it come from originally? And if he didn't write it, who did?

Chronology protection

The threat of paradoxes like these has prompted physicist Stephen Hawking to propose a 'chronology protection conjecture', which states that the laws of nature must prevent the creation of a time machine, thus preventing the occurrence of causality violations.

050 IN LIMBO

'Dammit, Briggs, I'm getting too old for this!' grumbled Murdo, wiping the sweat from his brow. Once again he was in extreme danger, and all thanks to his partner. Murdo was alone in a room with a diabolical device: a small bomb, triggered by a radioactive switch, which might go off at any moment. Briggs had been on the radio, trying to coach his partner through defusing the bomb, when there was a sudden crackle of interference. The radio link had gone dead.

'So he's in there solo, with no way of communicating with the outside world?' Briggs barked at the bomb disposal officer. 'Yes, sir, and the room is completely lead-lined, titanium-reinforced and electromagnetically shielded. There's no way in the universe to know what's going on in there until we get that door open.' 'Well, how long is that going to take?' 'Exactly 58 minutes.'

Briggs did a few calculations. He knew that the radioactive isotope in the trigger of the bomb was a single atom of nobelium-259, and that it had a half-life of precisely 58 minutes. This meant that in the time it would take to get the door open, there was a precisely 50 percent chance of the atom sending out a radioactive particle that would set off the bomb. And if that happened, Murdo would be dead. But there was also a 50 percent chance that he would still be alive.

Chief Randall arrived at the scene. 'I ought to take your badge for this, Briggs, but right now I'm more concerned with Murdo. Is he alive or dead?' Briggs scratched his head. 'It's complicated, Chief. You see, right now he's neither. Or both. Or maybe half one, and half the other. Technically, he's in a superposition of dual quantum states.' The Chief went bright red. 'Dammit, Briggs, I haven't got time for this. I want you to be the one who opens the door; everyone else will be outside the building in case of contamination.'

Watching Briggs enter the building, Chief Randall was struck by the thought that while Briggs might soon know whether his partner was alive or dead, as far as those waiting outside were concerned, Briggs himself would be in a superposition between relief and grief, an indeterminate state that would not be resolved until he came out or they went in.

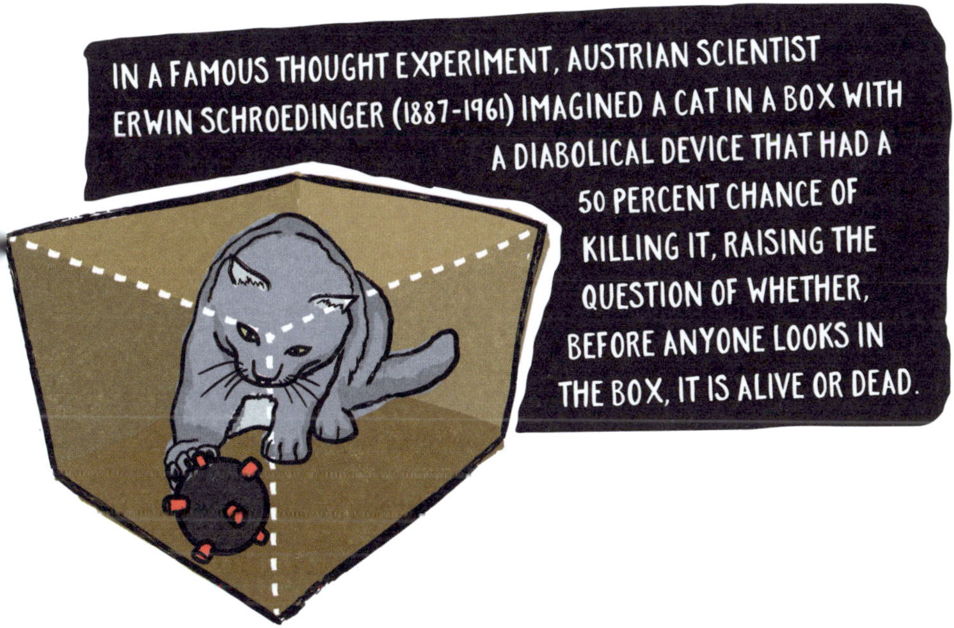

Schroedinger's ridiculous case

Murdo is like Erwin Schroedinger's famous cat, cited by the Austrian physicist as an illustration of the kind of 'ridiculous case' implied by the new quantum physics of the 1930s. According to quantum theory, some events are genuinely indeterminate until observed. The survival or death of the imaginary cat is one; it is not simply that the cat is one or the other but we just don't yet know which – the cat is literally neither one nor the other until observed.

051 JOURNEY TO THE CENTRE OF THE EARTH

Oliver Green proudly displayed to the assembled world media his latest and greatest futuristic magnum opus: the Trans-Earth Transport Tube (TETT). 'Thank you for joining me in Argentina', he began, 'where it is now possible to embark on the fastest and most remarkable transport link in history, and travel all the way to China'. He went on to explain the incredible challenges involved in digging a tunnel all the way through the Earth, passing directly through the centre of the planet. 'Our patented technology ensures the structural integrity of the TETT, even while a complete vacuum is maintained within, while our newly designed Travel Pods ensure the comfort of the passenger.'

Green explained that with a maximum acceleration of just 1G, the passenger experiences no more discomfort than when standing still in one place. As the pod falls toward the centre of the Earth, entirely powered by the planet's own gravity, with no need for engines, fuel or power supply, it reaches a speed of around 7,900m/s (17,700mph).

WHAT WOULD HAPPEN IF YOU FELL DOWN A HOLE THROUGH THE CENTRE OF THE EARTH?

'For a moment travellers will enjoy the unique sensation of weightlessness as they pass through the geometric centre of the planet, because at that point there will be no matter "below" you, but rather you will be surrounded equally in all directions by all the matter in the planet.' The inertia the pod develops as it falls carries it through the centre of the Earth and along the TETT towards the other end. As it travels, the pull of gravity on it increases, decelerating it at the same rate as it accelerated, so that its velocity reaches zero just as it pops up at the other end of the TETT in China. 'Once it reaches its destination,' Green explained, 'the pod is grabbed by robotic pincers, the door opens and the traveller exits without fuss. And the most astonishing thing about the trip?' he asked with rhetorical flourish. 'It takes just 38 minutes and 11 seconds!'

'Who will the first passenger be?' shouted a journalist. 'Why, I will, of course.' And with that, Green stepped inside the waiting pod, closed the door and pressed the release button. Almost immediately he realised that he'd made a fatal error in his calculations.

Shaken to the core

While Green has correctly calculated the speed and duration of a journey through the centre of the Earth, even managing to take into account the changing density of the planet as one moves from mantle to core, he has forgotten one rather important point: Earth is rotating, and the surface is moving at about 465m/s (1,040mph) relative to the centre. Anything at the surface partakes in this sideways motion, so that the pod is moving sideways relative to the TETT. As soon as it starts to drop, it will bash against the side of the TETT, and it will continue to rattle back and forth all the way down. By the time it comes back up the other side, assuming that the collisions are elastic, the pod will be travelling sideways at 930m/s (2,080mph, or three times the speed of sound), and the passenger is likely to be the worse for wear.

052 COLOUR ME BAD

Barillo and De Cecco were rival artists in Renaissance Florence, each famed for his creative use of colour and mastery of pigment. Lorenzo de' Medici was considering which of the two masters to choose for the most prestigious commission in Italy, a colossal landscape showing a rainbow over the city.

'I am the true champion of colour, Your Grace,' claimed Barillo, 'and I alone will achieve the right depiction of the rainbow.' 'Nonsense! Don't listen to this charlatan, Your Worship,' retorted De Cecco. 'I am the only one who knows the secret of the rainbow. Why, this fraud insists that there are eight colours in the rainbow!' 'Hah,' scoffed Barillo, 'and this imbecile actually believes there to be six.' At this point an ingenious young apprentice named Leonardo spoke up: 'Sire, I believe I may be able to tell which of the maestros is correct.'

Leonardo produced a piece of glass longer than it was wide, with a triangular cross section. 'This is called a prism, Your Eminence.' 'Impudent boy,' snarled De Cecco, 'His Grace knows as well as I do what a prism is, and that it adds colour to light by mixing it with darkness.' Barillo snorted: 'Everyone knows that.' But Leonardo persisted. He had the room made dark save for a narrow slit in one shutter, and placed the prism in the sliver of sunlight that came through. The beam of light spread out into a rainbow, the colours of which Leonardo highlighted by shining them onto white paper. 'You see,' exclaimed Barillo, 'just as I said – eight colours are clearly perceptible.' 'Poppycock,' retorted De Cecco. 'There are obviously six, anyone but a fool can see that.'

Lorenzo peered doubtfully at the spectrum of colours; he was not convinced it was at all obvious, and wondered if that meant he was a fool. 'Perhaps I might demonstrate something else, Your Eminence,' offered Leonardo, placing another prism in the path of the rainbow. The broad spectrum of colours shone on one side, and from the other emerged a beam of pure white light. 'And so you see, Your Eminence, that white light is actually composed of different colours of light, which can be separated and recombined.' Lorenzo thought that perhaps he had found the right man for the job.

ISAAC NEWTON, WHO FIRST EXPLAINED HOW RAINBOWS FORMED, SAID THERE WERE SEVEN COLOURS IN THE RAINBOW, BUT HOW DID HE COME UP WITH THIS NUMBER AND WAS HE RIGHT?

Unweaving the rainbow

In our story, Leonardo rather cheekily steals a momentous discovery from the future, a landmark demonstration by Isaac Newton, which Newton would develop in the late 1660s. The prevailing theory of colour held it to result from a mixture of light and shadow, black and white, but Newton neatly reversed this hypothesis with a pair of prisms. He showed that white light consists of a mixture of different colours of light, which refract to different degrees as they pass through a prism, so that sunlight can be separated into a spectrum or rainbow of colours.

The magic number

Newton claimed there were seven colours (red, orange, yellow, green, blue, indigo and violet), although in practice a spectrum has no clear boundaries between colours. His distinctions were arbitrary, reflecting his own occult obsessions, in which the number seven had specific symbolic significance.

053 MONKEY BUSINESS

Arthur closed the heavy door of the safe, turned the handle, spun the tumblers of the combination lock and declared, 'No one will ever discover the secrets lodged within.' Edith arched an eyebrow: 'As a matter of fact, I could break into that safe in less than three hours.' Arthur snorted, but Edith went on, 'The combination lock has only three tumblers, each bearing the numbers 0 to 9. This means there are only 1,000 possible combinations. If it takes me about 10 seconds to set one combination and try the handle, I could test all 1,000 in 10,000 seconds, which is about 167 minutes, or two hours and 47 minutes. And that's assuming that your combination is the last one I try.'

Arthur frowned and chewed his nails. 'But those are super-secret secrets in there! I'd better add another tumbler to the combination lock. Then no one will ever break in!' 'I wouldn't say "never",' chided Edith 'it would take longer, sure, but that's still only 10,000 possible combinations, or about 28 hours of trying combinations at most.' Arthur went away and bought a new lock for the safe.

'Right,' he said smugly, 'this new digital lock uses alphabetical characters so you can set a six-letter password. I'd like to see anyone get through that.' Edith hooked up her computer to the new lock and set it to work. 'There are 26 possible characters for each space in the password, so the "search space" – the total number of possible combinations – is 26^6, or about 309 million. It should take my computer less than a second to try all the possible combinations. There we go, your password is . . ."secret". Really, Arthur, I could have guessed that on my own.'

Arthur looked glum, so Edith took pity on him. She explained that the programme she had used was simply a 'brute force' password hacker, which goes through all possible combinations. Because computers work very fast, they can do this sort of task easily. 'The trick is to increase the size of the search space beyond the reach of brute force. Why don't you try using a few lines of poetry, or a passage you remember from a book? For instance, if you used "To be, or not to be" as your password, it would take my computer about 15,000 trillion centuries to crack it by brute force alone.' Arthur was not comforted: 'Hang on, I've heard all about this trick: Couldn't the computer just simulate a load of monkeys bashing at typewriters and work it out in no time?'

The infinite monkeys theorem

The British physicist Arthur Eddington (1882–1944) suggested that there is a greater chance of an army of monkeys accidentally typing 'all the books in the British Museum' than there is of a breach of the second law of thermodynamics. This has since become an illustration of both the power and the limitations of infinity. While it may be true that an infinite number of monkeys typing for an infinite length of time will eventually produce the complete works of Shakespeare, the universe is not infinite. According to one calculation, if the entire universe were filled with monkeys typing at random until the end of time, the chances of one producing Shakespeare's *Hamlet* would be approximately 1 in $10^{183,800}$.

054 REALITY BITES

'Aren't you coming to the big switch-on?' asked Cho. Kieu waved him away. She was too busy watching the Simpeople on her monitor. She found their little lives endlessly fascinating: The way they interacted like real people, fighting wars, falling in love, inventing new technologies, even trying to make art. All this despite being a relatively simple computer programme running on an old mainframe from the 21st century.

Cho didn't bother to ask again. He wasn't going to miss one of the most important moments in human history just because Kieu was obsessed with her stupid game. The Moon Computer was about to be unveiled, and he didn't want to miss it. Hopping on a shuttle, Cho rocketed into lunar orbit and joined the millions of others watching as the president turned on the greatest computer in history. Nanobots had converted the entire mass of Earth's satellite into a vast processing unit, capable of performing 10^{40} operations per second.

'People of the Solar System,' announced the president, 'today we begin a great project never before possible in the history of humanity. With the vast computer power now at our fingertips, we can simulate not just one genuine autonomous consciousness, but the entire mental history of mankind!

HOW DO WE KNOW WE ARE NOT LIVING IN A SIMULATED UNIVERSE, IN WHICH WE ARE COMPUTER-GENERATED VERSIONS OF REAL-WORLD CONSCIOUSNESSES, LIVING IN A VIRTUAL WORLD CREATED BY SUPER-ADVANCED REAL-WORLD CONSCIOUSNESSES?

This awe-inspiring technological achievement will allow us to model our own cultural and intellectual evolution, and gain extraordinary new insights as we seek to extend our powers still further.'

Back on Earth, Kieu watched her Simpeople in amazement. Some of them had stumbled upon the principles of computing, and were even now designing their own digital computer. The read-out monitoring the parameters of her simulation started to flash red, warning that the antiquated mainframe processors were nearing overload. Clearly the system could not cope with the processing demands of a computer within a computer. Regretfully she terminated the simulation, and turned her attention to the news. But when she heard the president's address, she was struck by a terrifying revelation. In horror she activated the comms link.

'Cho! Cho! You must stop the switch-on,' Kieu screamed down the link. 'If it goes ahead, we'll . . . we'll be turned off.'

The simulation hypothesis

Philosopher Nick Bostrom (b. 1973) posits that if humanity survives into the far future, it will probably achieve immense computational power, and that this might be used to create simulations featuring fully functional human-like consciousnesses. If real minds can simulate entire universes of minds, the chances are that they will run many such simulations, in which case simulated minds will vastly outnumber real ones – which means you are vastly more likely to be one of the simulated ones.

Information overload

In our fictional scenario, Kieu has just realised a risk of which Bostrom has warned. If we are living in a simulation, and we advance to the point where we can run our own fantastically complex simulations, it is possible that we would thus exceed the computational constraints of the simulators, who would then be forced to terminate their simulation.

055 CHEATING DEATH

Nkwambe is eating a sandwich when Death comes. Nkwambe stops eating her sandwich. 'Is it the sandwich that kills me?' she asks Death. His head, or at least the dark void covered by his black hood, rotates slowly, first one way, then another. 'I'll take that as a "no". Hmm, I wonder what it can be? Surely I'm too young to have a heart attack.' Death extends a skeletal hand and tilts it from side to side. 'Anyway,' Nkwambe continues, 'it doesn't matter. I'm not ready to go. Let's play a game. I know you like those. If I win, I get to live, and if I lose I'll come quietly.' 'VERY WELL,' intones Death. 'WILL IT BE CHESS, OR PERHAPS POKER?'

Nkwambe suggests that they flip a coin, not just once but 20,000 times. 'YOU ARE PLAYING FOR TIME. BUT IT WILL NOT AVAIL YOU – ALL TIME IS INSTANTANEOUS TO ME.' Death produces a large golden coin, engraved on one side with a serpent's tail, and embossed on the other with a skull and crossbones. 'HEADS OR TAILS?' Nkwambe thinks fast. Death has laid the coin on his bony thumb with the head face up. She reasons that, no matter how many times it flips once tossed, since it started heads up it must be fractionally more likely to land that way, for the same reason that if you count numbers starting with an odd number you will

A COIN TOSS IS THOUGHT OF AS THE EPITOME OF RANDOMNESS, YET STUDIES HAVE THROWN INTO DOUBT THE IMPARTIALITY OF THE COIN. WHY ARE COINS UNFAIR?

always have counted either the same or a greater number of odd numbers than even. 'YOU ARE THINKING THAT STARTING WITH HEADS MEANS THAT, ON AGGREGATE, THE SPINNING COIN WILL SPEND MORE TIME HEADS UP THAN TAILS. YOUR REASONING IS FAULTY.'

So Nkwambe makes an alternative suggestion: 'Let's spin the coin instead.' Death looks uncomfortable. 'I DO NOT LIKE THAT IDEA. THIS COIN IS CLEARLY HEAVIER ON ONE SIDE THAN THE OTHER. THE HEAVIER FACE IS MORE LIKELY TO END UP ON THE BOTTOM.' 'Fine, we'll stick with the flip,' says Nkwambe confidently. 'I choose heads.' She wins her bet.

Probably wobbly

Nkwambe knows something that Death does not. Because of precession – the way that a coin wobbles on its axis as it flips through the air – there is a very slightly increased probability that a coin will land (assuming it doesn't bounce) with the same face up as when it started. After 20,000 flips starting from a heads-up position, Nkwambe can expect roughly 10,200 heads, although she will be in trouble if Death is one of those people who catch a coin and then turn it over onto the back of the other hand to call it.

In a spin

Death is right about spinning a coin. For instance, an American penny is heavier on the side with Lincoln's head than on the reverse, 'tails' side, which features the Lincoln Memorial. Spin one of these coins and it will land tails side up roughly 80 percent of the time.

056 WHAT GOES UP

Bev the Baker is famous for the remarkable speed at which she bakes pies. Gluttonous Gus is famous for his appetite for pies. Bev reckons she can make a pie as fast as Gus can eat it. Mayor Meyer times each of them and confirms it: Bev can bake 10 pies a minute, while Gus can eat 10 pies a minute. Gamblin' Gavin proposes a contest: Bev will be given the ingredients to make 100 pies. If they disappear in less time than it took to bake them, Gus will be the winner. Gamblin' Gavin says that he will take bets on the outcome, and that in the event of a draw he will pay out to all bettors.

Mayor Meyer thinks that Gavin is crazy. Simple arithmetic says that Gus will take the same amount of time to eat the pies as Bev took to make them. Gavin points out that Mayor Meyer has forgotten the mice that plague their town, and which constantly nibble at any pastry that is left sitting around. Meyer admits there are mice but insists that they cannot really affect the outcome because it takes them 10 minutes to eat a single pie. Gavin says this will be quite enough to ensure that his bets on Gluttonous Gus pay off.

The contest starts. Bev bakes like a woman possessed, but although she works at top speed the mice nibble away at her pies. In the 10 minutes it takes her to make 100 pies, the mice have eaten one and so she has to spend another 6 seconds making a replacement. As soon as she has finished, Gus gets to work. The mice are also eating away, consuming another whole pie and thus leaving only 99 for Gus to guzzle. The pies are all gone in 9 minutes 54 seconds. Gus has won the contest by 12 seconds. Gavin gets rich.

'The most impressive fact is that gravity is simple . . . and therefore it is beautiful.'

– Richard Feynman (1918–88)

Joining forces

Gus and Bev are analogous to a ball going up and down respectively. The rate at which they can eat or make pies is analogous to the acceleration an object experiences in a gravitational field. The mice are analogous to air resistance. The hundred pies get eaten more quickly than Bev can make them, even though both she and Gavin work at the exact same rate, because the mice are helping him and hindering her.

A ball going up experiences both gravity and air resistance acting in the same direction, working together to decelerate it from its initial speed to zero. A falling ball experiences the force of air resistance acting in the opposite direction to gravity, working against each other as the ball accelerates from zero. Hence it takes the falling ball longer to cover the same distance as the rising one.

A BALL EXPERIENCES THE SAME FORCE OF GRAVITY ON THE WAY UP AS ON THE WAY DOWN, SO WHY SHOULD IT TAKE LONGER FOR A BALL TO COME DOWN THAN GO UP?

057 AGE-RELATED DECLINE

Mrs Sharpe poured herself another cup of tea and sat sipping it, peering at the other guests over the rims of her half-moon spectacles. They stirred uneasily. 'Shall we recap what we know so far? Poor Colonel Arbuthnot was murdered in London just eight days ago, but all of you ladies say you have been here in Dorset for more than two weeks. That would rather seem to rule you out, wouldn't you say?' The ladies relaxed, but not for long.

'I have, however, noticed that none of the four of you is a real brunette, although you all share the exact same taste in hair dye. I also know that this particular shade of brown can only be obtained at Monsieur Bruin's in Mayfair. So I wonder if you might indulge a little fancy of mine, and be so good as to provide a lock of your hair to my friend Sir Henry, who will take it to be analysed?' The four ladies grumbled but consented.

Two days later, Sir Henry brought Mrs Sharpe the results of the analysis. 'Well, old friend, it is much as I expected. It seems we have our suspect.'

HOW CAN WE MEASURE THE AGE OF ARCHAEOLOGICAL REMAINS, AND WHAT CAN SUCH TECHNIQUES TELL US ABOUT THE AGE OF THE TURIN SHROUD?

Sir Henry was most impressed; how could she be so sure? 'These analyses show that in three of the women the ratio of brown hair to their real colour is less than nine to one. Assuming that all three ladies have hair that grows at the same rate, this means that none of them has had their hair dyed brown within the last two weeks. Mrs Stryker, on the other hand, has hair that is 99 percent brown, proving that she must have had it dyed within the last week, and thus that she must have been in London much more recently than she claims!'

Carbon dating

Mrs Sharpe is using the way that the ratio of dyed to undyed hair decreases over time to date the last visit of the suspects to London. Archaeologists can use a similar principle to date organic remains, such as animal bones, charred wood or ancient fabrics. A minute but relatively constant proportion of the carbon atoms in atmospheric carbon dioxide are radioactive carbon-14 (C14). While they are alive, plants continually cycle fresh carbon from the atmosphere through their tissues, so that they contain the same proportion of C14 as the atmosphere, as do animals that feed on the plants. Being radioactive, C14 decays into the much more common C12 isotope, and as soon as an organism dies and stops replenishing its store of C14, the ratio of C14 to C12 will begin to decline. Measuring how much it has declined – known as carbon dating – gives a measure of how long organic material has been dead.

The Turin Shroud

The Turin Shroud is claimed to be the burial shroud of Jesus Christ and to contain a miraculous image of his crucified body. Carbon dating of samples of the fabric show that the ratio of C14 to C12 is consistent with what would be expected from plant material harvested in the 14th century. Critics of the dating studies argue that the fabric tested came from patches used to repair the Shroud, but it seems telling that the first undisputed record of its existence dates to around 1350.

058 UNEASY RIDER

Captain Svenquist surveyed the strangely scalloped surface of Planet Kepler-B238 with dismay. She was to be stationed here for a year, and had counted on spending her downtime pursuing her favourite pastime – biking. It hadn't been easy getting the Intergalactic Authority to authorise bringing a bicycle on the spaceship, and having brought her much-loved mountain bike across 400 light years she was damned if it was going to sit idle.

She got down on her hands and knees and had a close look at the unusual ground formation. The surface of the planet was rippled like a frozen seascape, but with remarkable uniformity. Her hand scanner showed that the rounded ridges and troughs of the ripples followed a mathematically precise form, no doubt reflecting the unique weather conditions of Kepler-B238. According to the read-out, the cross section of the ripples revealed a sequence of perfect inverted catenary curves.

Undaunted, Svenquist set up the bike and started to ride, but she didn't get further than a single revolution before the bone-juddering reverberations forced her to stop. Clearly the bike would be shaken to bits in seconds. There must be

HOW CAN YOU RIDE A BIKE WITH SQUARE WHEELS?

some way to get a smooth, level ride. What was needed was a shape where each corner would describe an inverted catenary as it rotated around its centre. A moment on the computer revealed the answer: a square. Svenquist got to work with the 3-D printer and in minutes she was swapping out her circular wheels for square ones. Setting out across the rippled alien landscape, she was delighted to feel a smooth, level motion without the slightest bump or deviation – at least until she tried to turn the wheel . . .

Smooth operator

The corner of a square rolling around its centre describes an inverted catenary curve as it moves through space. This enables square wheels to roll over a bed of inverted catenaries of appropriate size, with the axles of the wheels staying level and moving at constant speed. Mathematician Stan Wagon (b. 1951) of Macalester College in St. Paul, Minnesota, has a square-wheeled tricycle that he rides over a course of inverted catenaries, but it only works when going in a straight line.

A wheel for all surfaces

Different shapes describe different curves or figures as they roll around their axes. There is a wheel shape that can achieve a smooth ride across almost any geometrically regular surface. For instance, a sawtooth zigzag surface can be traversed smoothly with a flower-shape wheel composed of four 'petals' made from sections of equiangular spirals.

'Life is like riding a bicycle. To keep your balance you must keep moving.'

– Albert Einstein (1879–1955)

059 WHO'S SCRYING NOW?

Werner Heisenberg and Lord Kelvin brushed aside the bead curtain and entered the parlor of Madame Avenir, Prognosticator to the People (according to the sign in her window). There was a heady scent of incense, and red cloths covered all the lamps. Madame Avenir gestured for the eminent physicists to sit down. 'We understand that you claim to be able to divine the future, Madame,' began Lord Kelvin, 'and we have come to put a stop to your nonsense.' Madame Avenir raised an eyebrow, and then placed a small box on the table. Out of the box climbed a little imp holding a calculator. 'Get to work,' she told the imp, who started frantically bashing the keys of his calculator. 'Soon he will tell us the outcome of this meeting,' she explained.

'Not so fast, Madame,' exclaimed Heisenberg, producing a small vial of radioactive material and a Geiger counter. 'Let him predict the next blip on the counter, if he is able.' The imp looked flummoxed. Lord Kelvin joined in, placing on the table a container of gas. 'I challenge your creature to predict the distribution of every particle of gas in this box in five minutes' time.' The imp began to sweat. Heisenberg

IF THE FUTURE STATE OF THE UNIVERSE IS DETERMINED BY ITS PAST AND PRESENT STATES, THEN COULDN'T A BEING WITH ENOUGH INFORMATION USE THE LAWS OF PHYSICS TO DETERMINE THE ENTIRE HISTORY OF THE UNIVERSE?

delivered the killer blow, pulling from his pocket a small jar containing a butterfly and releasing it into the room. 'Let him tell us what effect this butterfly's wing beats will have on the weather in Florida over the next few months.' The imp exploded in a puff of purple smoke, leaving behind a small piece of paper. Madame Avenir read what was written on it: 'He predicted you would succeed in destroying him, and then spend the rest of the day listening to the radio.' Heisenberg and Kelvin looked at each other and smiled. 'I believe we'll go to the cinema instead.'

Laplace's demon

In the Enlightenment era there was a growing belief that the universe is essentially deterministic, which means that its future states can be predictably determined, given enough knowledge about its current and past states. Eminent thinkers from Cicero (106–43 BCE) to Leibniz (1646–1716) to Laplace (1749–1827) proposed that a being sufficiently well-informed, and with enough processing power, would thus be able to determine the whole history of the universe, past, present and future. This conjectural being has become known as Laplace's demon. Madame Avenir employs the services of just such a demon, but discoveries in nineteenth- and twentieth-century physics banish it.

Uncertainty principles

Laplace's demon is destroyed by a number of principles. The Second Law of Thermodynamics shows that information is not conserved but inexorably lost. The uncertainty principle in quantum physics shows that it is impossible to determine precisely both the position and the momentum of a particle at the same time, and makes radioactive decay a truly random process. Chaos theory in mathematics means that increasing precision of measurement, which the demon depends upon, does not produce increasing precision of determination.

060 AROUND THE WORLD

Thenes lives on a small planetoid orbiting a Sun-like star. He is planning a circumpolar navigation of the planet and needs to know how much fuel to put into his land cruiser, but no one can tell him until he meets Eratos, who thinks she can help. On Midsummer's Day she goes to the town of Aaa, where the sun is directly overhead, and sends Thenes to the town of Bbb, 100 units north. They each plant a pole in the ground so that it sticks straight up in the air. When the Sun is directly overhead at Aaa, so that the Aaa pole casts no shadow, Eratos signals Thenes to measure the angle between the top of his pole and the tip of its shadow.

Eratos now has all the information she needs to work out the circumference of the planetoid, with the help of some mathematical laws she learned from an ancient Earth geometer called Euclid (c. 325–265 BCE). The first law is that a line drawn perpendicular to a tangent with a circle points directly to the centre. In this case, lines from the two poles to the centre of the planetoid are the perpendiculars, which define a slice of the circle of the planetoid. If Eratos can work out the angle of that slice, then she can work out the proportion of the circumference of the planetoid represented by the distance between Aaa and Bbb on the surface.

HOW DO WE KNOW THE EARTH ISN'T FLAT? MORE TO THE POINT, IS THERE A WAY WE CAN PROVE IT MATHEMATICALLY AND GEOMETRICALLY WITHOUT GOING UP INTO SPACE?

To work out this angle, she uses a second Euclidean law, which says that when a line crosses parallel straight lines, the alternate angles (the angles inside the parallel lines, on either side of the crossing line) are equal. The parallel straight lines here are the rays of the sun, and the crossing line is the perpendicular line from Bbb to the centre of the planetoid. The angle that Thenes has measured at Bbb is equal to the angle of the slice of the circle described by the three points Aaa–centre–Bbb. Thenes measured 6°, so that slice also has an angle of 6°, which is equivalent to 1/60th of a circle (there being 360 degrees in a circle). Since the circumference of the slice is 100, the circumference of the whole planetoid must be 6,000 units. Secure in the knowledge that he has enough fuel to travel 6,000 units, Thenes sets off on his circumpolar voyage.

High noon in Syene

Eratos has duplicated a landmark finding by the ancient Hellenic scholar Eratosthenes (c. 276–194 BCE), who was in charge of the library at Alexandria in the third century BCE. Eratosthenes had heard that at noon on the summer solstice the sun cast its rays all the way to the bottom of wells in the Egyptian town of Syene, and that a staff stuck in the ground cast no shadow. He knew that at this exact same time of day, a staff stuck in the ground in Alexandria *did* cast a shadow. Using this information and some simple geometry, Eratosthenes proved that the Earth was spherical, and calculated the circumference of that sphere to within 1–2 percent of modern values.

'[Eratosthenes] . . . is a mathematician among geographers, and yet a geographer among mathematicians . . .'

– H. L. Jones (ed.), *The Geography of Strabo* (1917)

061 FIZZLING OUT

McBlammo jumped down from the helicopter and surveyed the wreckage of the tank. Dr Evil struggled to free himself from the twisted metal of the cabin, his henchmen pushing and shoving behind him. With grim satisfaction McBlammo put a cigarette in his mouth, lit it, took a deep drag and exhaled. 'They say smokin's bad for you. I guess there might be something in that,' he drawled, flicking the glowing end of the cigarette into the spreading pool of petrol.

'CUT!' screamed the director, gesturing wildly at a man in a white lab coat who had wandered into the middle of the frame. 'Who the hell is that, and how did he get on my set?' The man in the white lab coat peered disapprovingly at the now soggy cigarette. He flashed a business card that read 'Science Inspector'. 'What the hell is a science inspector?' the director ranted. 'I'm sorry, but I'm going to have to shut you down. Your film simply contains too many scientific fallacies.' 'Falla . . . what now?'

'Take this petrol stunt, for instance. It's total nonsense. You cannot light liquid petrol with a cigarette.' The actor playing McBlammo, Scotty Francheiz, protested. 'Why not? The end of a lit cigarette burns at 900°C (1,652°F); that should be more than enough to set fire to petrol.' 'Yes,' added the actor playing Dr Evil,' and petrol vapour ignites at around 250°C (482°F).' The science inspector shook his head. 'But you are not

IT'S A STAPLE PLOT DEVELOPMENT IN MANY A BLOCKBUSTER FILM, BUT CAN YOU REALLY SET FIRE TO PETROL BY DROPPING A MATCH ON IT?

lighting vapour – you threw the cigarette directly into the liquid petrol, which simply extinguished it. And even if you held the glowing cigarette in the vapour above the petrol, it would not be able to transfer its heat fast enough to make the vapour ignite.'

'Listen, you nerds,' roared the director, 'this film is on a tight budget, and time is money. I can't afford to listen to this rubbish. What's it gonna take to get this petrol lit?' Scotty Francheiz piped up: 'How about a naked flame? That would set off the vapour, and then the flames would spread to the liquid petrol.' The science inspector nodded, and was about to warn of the extreme explosive dangers of doing so when the impatient director lit a cigarette lighter and held it over the pool of petrol, declaring, 'One way or another, this picture's gonna end with a bang!'

Hot enough?

The science inspector is right: Its vapour is the truly flammable element of petrol, which makes it so dangerous. But even lighting vapour turns out to be a complex business. According to a 2010 study by forensic scientist Rebecca Jewell *et al.*, the dynamics of heat transfer from a hot surface (such as a lit cigarette end) to petrol vapour are such that the surface needs to be at 980–1130°C (1,796–2,066°F) before ignition will occur; and a cigarette never gets quite this hot.

Power-packed

Experimenting at home with petrol is incredibly dangerous and should never be attempted. Setting fire to just one-third of a cup of petrol releases as much energy as striking 5,100 matches at once.

062 FREE FALL

Einstein and Newton are trapped in a falling elevator. They debate how best to survive the drop. Newton ponders relative motion. Perhaps, he suggests, a powerful leap at the last instant before impact will cause an upward acceleration counter to the downward acceleration of the falling lift, thus ameliorating if not cancelling out the tremendous force of impact. Einstein points out potential flaws in this reasoning. Even the maximum force either of them could generate would be trivial compared to the tremendous momentum generated by their uncontrolled descent, and this is probably for the best since otherwise they would smash their heads on the roof of the elevator.

Also, he points out, in order to jump they would need to push off from the floor, but neither of them had their knees bent when they started to drop. Now that they are in free fall, if they bend their knees their feet will simply move up off the floor. Einstein suggests that the wisest course would be to lie down, since in this fashion the impact force will be spread out over a wider area, and the most important bones will be horizontal to the impact rather than perpendicular.

COULD YOU SURVIVE IN A FALLING ELEVATOR BY JUMPING AS IT HIT THE GROUND?

Newton counters that since they are in free fall they cannot simply drop to the floor, but will have to pull themselves down using the sides of the elevator and then hold themselves there. He also wonders whether Einstein has considered the important variables in determining the acceleration their bodies will undergo when the elevator hits bottom. They agree that Newton will stay upright and try to jump, by flexing his toes if necessary, while Einstein will use the rail to push himself to the floor and lie as flat as possible. Who will fare better on impact?

Bend the knees

Advice to unfortunates who find themselves in plunging elevators differs: Many sources advocate the lying flat option, but Newton has a point. The acceleration (or rather, rapid deceleration) experienced by the brain on impact is what will do the damage, so the greater the acceleration rate, the worse off the faller will be. The magnitude of the acceleration (a) is given by the equation $a=v^2/2d$, where v is the velocity on impact and d is the distance travelled during stopping. The bigger d is, the smaller a will be. So the extra 1.5 metres or so gained by standing upright (providing the knees are flexed to give some degree of 'crumple zone') could make a slight difference to the chances of your brain surviving.

Record breaker

The holder of the record for longest elevator fall survived is Betty Lou Oliver, who survived falling 75 stories (more than 300m/1,000 feet) in an elevator in the Empire State Building, after it had been hit by an aircraft in 1945. It is thought that an air pocket or the coiled elevator cable itself broke the fall.

063 THE FALL

Almost every morning at breakfast, Xander curses his clumsiness. He takes his toast from the toaster, puts it on a plate, spreads it thickly with butter (his personal vice), and carries it to the kitchen table. Unfortunately there always seems to be something in the way, and Xander always trips over it. Worst of all, the toast invariably, inevitably lands butter-side down on the floor. When Xander retrieves it, he finds it covered in cat hair and kitchen-floor gunk, and he has to start again from scratch.

Determined to break the pattern, Xander tries all sorts of variations. He buys different types of bread, cuts his own slices from a loaf, spreads different preserves, tries out different butter and margarine spreads, toasts the bread for varying lengths of time – all to no avail. It seems to make no difference whether he spreads a paper-thin layer of butter, or slaps on a slab; the toast ends up butter-side down on the floor every time.

Finally, in despair, Xander appeals to his wife to help. She waits until he has buttered his toast, then turns it upside down. He still drops it on the floor a moment later, but this time it lands butter-side up.

'The chance of the bread falling with the buttered side down is directly proportional to the cost of the carpet.'

– **Arthur Bloch, *Murphy's Law* (1977)**

The butter end

Experiments with thousands of trials prove that toast really does tend to land butter-side down when it falls on the floor – or more accurately, that whichever side was facing up to start with will end up facing down when it lands. This is nothing to do with the weight of the butter or the aerodynamics of the toast, but simply a function of the height from which toast is usually dropped. As it falls, toast will flip end over end. If it falls from high enough, the toast will flip around 360 degrees and land butter-side up, but if not, there will only be time for a half-rotation, meaning it will land butter-side down. Experiments show that, given the typical speed of rotation of dropped toast, the minimum height needed for a safe landing is about 2.4m (8 feet). Since few people butter or eat their toast on surfaces this high off the floor, the odds greatly favour the messier outcome. One suggestion for improving the odds is to press hard when buttering, to make the toast bow upward at the sides.

064 LEAN IN

Blessing has been given a boomerang for her birthday. She has read all about boomerangs, and how the Australian Aborigines used boomerangs for hunting, and she has hazy pictures in her mind's eye of a boomerang somehow rounding up kangaroos and bringing them back to an Aboriginal hunter. She goes to the park to try out her gift, but it proves to be a lot more difficult to use than she had imagined.

At first she tries to throw it like a frisbee, holding it horizontally and tossing it backhand. The boomerang falls to the ground. Then she tries a similar throw, but forehand. The boomerang falls to the ground again. This isn't how she imagined it at all. A passing lady kindly shows her the correct way to throw it: by holding one end, tilting the boomerang so that it is almost vertical, and throwing it end over end. Blessing throws it the way she has been shown, and to her delight it hurtles around in a wide circle and comes to rest on the ground not far from her feet.

Blessing is filled with a sense of triumph, but she is still left with two questions. How does the boomerang manage to fly in a circle, and how can it round up kangaroos while it is doing so?

WHAT ARE THE PHYSICS THAT MAKE A BOOMERANG TRAVEL AROUND IN A CIRCLE AND END UP BACK WHERE IT STARTED?

Hunting weapon or toy?

Blessing's confusion about hunting with boomerangs is a common misconception. There are two types of boomerang. One is a weighted throwing stick that Aborigines used to hurl at prey animals with deadly force, intended to fly straight and true. The returning boomerang has been used by Aborigines as a toy.

Gyroscopic precession

A rotating boomerang is like a gyroscope. Pushing on one side of a rotating wheel or gyroscope causes a force to act perpendicular to the direction of your push. This is called gyroscopic precession. Push the top of a rotating wheel away from you, and instead of toppling over it will turn away from you. It's how you can steer a bike without the handlebars, simply by leaning. So generating some sort of sideways force or push acting on the top or bottom of a boomerang can make it turn away from its straight-line path, and keeping the force applied will turn it right around in a circle.

In a spin

The two arms of a traditional returning boomerang are shaped like two aeroplane wings stuck together. As it both spins and travels through the air, the top arm generates more lift than the bottom one, since its speed through the air is the sum of the speed of travel plus the speed of rotation, whereas the speed of the bottom arm is the speed of travel minus the speed of rotation. The faster a wing moves through the air, the more lift it generates, so more lift is generated by the top arm than the bottom one, causing a disparity that produces a net sideways force. Since the spinning boomerang is like a gyroscope, this sideways force is translated into a force perpendicular to the spin, causing the whole boomerang to turn or precess.

065 REVOLUTIONS

Luca is a geocentrist: He believes the Earth is the centre of the universe, and the Sun goes around it once a day. You are trying to convince him that the Earth is actually spinning, and this is why the Sun seems to cross the sky every day.

Luca thinks this is ridiculous, and he explains why. If your contention is correct, the Earth must be rotating at an incredible speed. More specifically, he says, while a person at one of the poles might simply be turning around once a day, someone at the equator would have to be hurtling through space at a quite fantastic rate.

Imagine a traditional record player, he says. The inner circle through which the spindle passes is only around 3cm (1.8 inches) in circumference, he estimates. Turning at a rate of 33rpm, a tiny person standing on the inner edge of a record must therefore be travelling 99cm (39 inches) every minute, or about 0.06km/h (0.037mph). The diameter of an LP is 30cm (11¾ inches), so its circumference must be about 94cm (37 inches), yet the outer edge also makes a complete circle 33 times a minute. This means that a tiny person standing on the outer edge of the record must be travelling 3,102cm (1,221 inches) every minute, or roughly 1.9km/h (1.2mph).

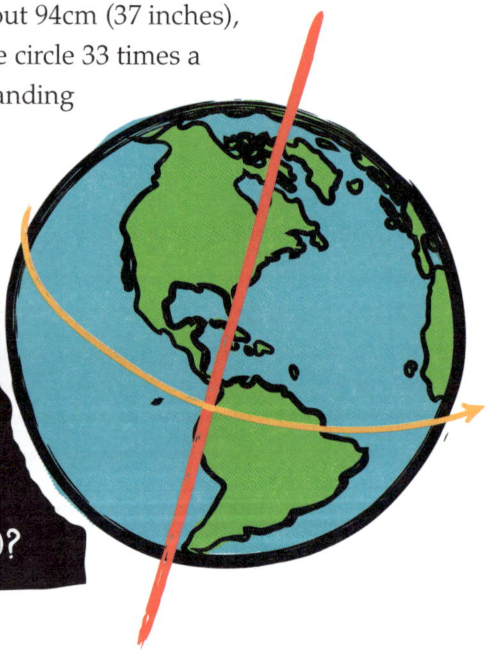

THE EARTH IS PRETTY BIG, AND IF IT IS ROTATING ONCE EVERY 24 HOURS IT MUST BE SPINNING EXTREMELY RAPIDLY. JUST HOW FAST IS THE EARTH GOING AROUND?

If the Earth really is rotating about its axis once a day, he argues, someone standing at the equator would be like the little man on the edge of the record. We know, Luca says, that the Earth is about 12,756km (7,926 miles) across. Using the formula $2\pi r$, this means that its circumference at the equator must be around 40,074km (24,900 miles), meaning that the equator – and everything and everyone on it – must be moving at a speed of 1670km/h (1,038mph).

This is patently ridiculous, Luca points out, since if the Earth were really hurtling around at high speed we would all be buffeted by jet-force winds, and anyone who jumped into the air would find that the Earth's surface had rotated several yards under their feet before they come back down.

The Coriolis effect

To dispel Luca's fallacy you would need to explain about momentum, and how everything in contact with the Earth – including people, and the atmosphere – shares its momentum. You could also point to a number of real and measurable effects of the rapid rotation of the Earth. For instance, the Earth's rotation causes the oceans to pile up on their western edges – sea level on the western shores of the Pacific is 45cm (18 inches) higher than on the eastern shores. It also causes the Coriolis effect, where things moving from high to low latitudes appear to be deflected east or west because their momentum differs from that of the region to which they are moving. This is why weather systems appear deflected to the west as they move from the poles to the equator. A naval shell fired from north to south, covering 24km (15 miles), is deflected by around 90m (295 feet) from its 'straight line' destination.

'No one ever yet felt or saw the earth careering through space at the terrific rates it is credited with . . .'

Thomas Winship (1899)

IN SPACE

'Astronomy compels the soul to look upwards, and leads us from this world to another.'

– Plato, *The Republic* (342 BCE)

066 WHEN THE SUN GOES DOWN

In astronomy class, Anander, Cho and Mara are learning about the future of the Sun. The Sun, the lecturer tells them, is relatively stable, but as it burns its hydrogen into helium, the heavier helium accumulates in its core. This raises the density of the core, increasing the compression due to gravity, raising the pressure and intensifying the nuclear fusion reactions. Gradually the Sun burns hotter and brighter; it has increased in luminosity (the measure of its energy output) 30 percent since its birth, 4.5 billion years ago. Over the next billion years it will increase another 10 percent in luminosity, driving a runaway greenhouse effect on Earth that will transform our planet into a second Venus.

But, they learn, this will only be the start of the Earth's problems. In about 5.4 billion years the Sun will balloon into a red giant, expanding beyond the orbit of Venus, while its tenuous outermost atmosphere

WHEN THE SUN NEARS THE END OF ITS LIFE, IT WILL BURN HOTTER AND HOTTER, AND WILL EVENTUALLY EXPAND INTO A RED GIANT BIGGER THAN THE ORBIT OF VENUS. WHEN THIS HAPPENS, IS THERE ANY WAY THE EARTH COULD BE SAVED?

will capture the Earth, dragging it closer, until it plunges into the fiery maw of the giant. The Sun itself will eventually (in just under 8 billion years from now) blow off its outer layers, leaving behind a white-hot, incredibly dense mass of carbon, slowly cooling in interstellar space.

Having presented the threat to future Earth, the lecturer now tasks her students with working out how to save the planet. Anander's idea is to construct a large ion drive on Earth, to be aimed up into the atmosphere, and fired only when pointing directly at the Sun. As ions are ejected from one end of the drive, there should be an equal and opposite reaction, pushing the drive down against the ground, and, by extension, pushing the Earth away from the Sun.

Cho's plan is to send small robots into space to capture asteroids and shift them closer to the Earth. The tiny gravitational attraction of each asteroid should effect even tinier perturbations in the orbit of the Earth, pulling it further away from the Sun. Mara's idea is to set up a vast solar sail, floating far out in space just within the orbit of the Moon, attached to the Earth by fantastically strong nanotube filaments linked to huge anchors at the equator. The sail would use the light pressure of solar photons to move the planet into a more distant orbit, like a celestial kite surfer.

Will the Earth move?

Even if there were no atmosphere to interfere with the passage of the ions and soak up their momentum, it seems likely that Anander's plan would simply take too long to be effective. Cho's idea is technically feasible (or at least should be in the future) and theoretically sound, but might accidentally set an asteroid on a direct collision course with Earth. Mara's idea is currently far beyond the limits of technology, and would require the development of new materials. Ideally the rate at which Earth's orbit would be shifted further from the Sun would be made to balance with the rate at which its luminosity is set to increase, to ensure a relatively constant intensity of solar radiation hitting the Earth.

067 IS THERE LIFE ON MARS?

At a meeting of the Society for Interplanetary Biology, three scientists present discoveries they claim to have made on Mars. Each claims to have discovered life! Professor A displays the carcass of a large six-legged animal with batlike wings. Professor B displays a fleshy plant similar to a cactus, which he claims to have picked from the summit of the Martian volcano Olympus Mons. Professor C displays a slide, on which can be viewed with the aid of a microscope a tiny single-celled organism with thick cell walls, which closely resembles an Earth bacterium. He explains that it was retrieved by drilling deep into the Martian crust. The learned members of the Society are not permitted to touch, probe, dissect, test or even look too closely at the supposed life forms, yet they are expected to judge which are likely to be real, and which bogus.

PHILOSOPHERS AND SCIENTISTS SPECULATED ABOUT LIFE ON MARS UP UNTIL THE TWENTIETH CENTURY, AND IT WAS BELIEVED THAT MARTIANS MIGHT EVEN HAVE BUILT A CANAL SYSTEM. HOW HAVE RECENT DISCOVERIES ABOUT MARS CHANGED OUR EXPECTATIONS?

Cold comfort

Thanks to its tectonic activity and weathering, Earth maintains a relatively constant level of carbon dioxide in the atmosphere, creating a greenhouse effect without which the surface temperature of our planet would be –18°C (0°F) instead of around 15°C (60°F) on average. Mars's lower gravity has allowed much of its atmosphere to drift away; lack of tectonic activity means that once atmospheric carbon is locked away as carbonates due to weathering, or as snow when carbon dioxide freezes, it stays put and is not recycled into the atmosphere. These factors make the Martian atmosphere 100 times thinner than Earth's, and so there is no greenhouse effect and the surface is around –55°C (–67°F) on average. Any surface water would freeze, and even if it didn't, the low pressure would make it evaporate.

The lack of atmosphere allows intense radiation to hit the surface of Mars. Any organisms on the surface would be subjected to UV radiation that would scramble their molecules – one of the most serious hurdles facing a crewed mission to Mars.

Picking a winner

Bearing in mind the immense challenges facing potential Martian life forms, the Society for Interplanetary Biology decides that two of the purported finds are hoaxes. Professor A's land animal would not survive five minutes on the surface of Mars, let alone a lifetime of being blasted with ionising radiation. Professor B's plant could not withstand the radiation, intense cold and near-vacuum atop the enormous Olympus Mons, which has just 8 percent of normal Martian atmospheric pressure. Professor C's microbe is much more plausible: If there is life on Mars, it is likely to be found underground, away from the radiation and freezing, dessicating conditions. Earth has microbes that survive in extreme conditions such as deep underground, and any life found on Mars may well resemble these 'extremophiles'.

068 MOON REAVER

Al finds a magic wishing ring on a local beach. He decides to get a good night's sleep before deciding what to wish for. Unfortunately there is a full moon and its bright light stops him from dropping off. 'I wish that stupid Moon had never existed', he says out loud, somewhat carelessly.

Four billion years earlier, a Mars-sized planetoid is hurtling towards the newborn Earth, on track to collide with it and get smashed into a cloud of debris, which will coalesce into a new satellite for the now slightly battered Earth. At the last moment the planetoid changes course, missing the Earth and plunging into the Sun. The young Earth continues to spin at its previous, extremely rapid rate, so that each day lasts just 5 hours. Over the next 4 billion years it doesn't slow down that much.

Al is awoken by a number of strange things. He sees the Sun moving rapidly across the sky, and since he lives at a high latitude the atmosphere is dramatically thinner than usual, and the sunlight more intense and dangerous.

WHAT HAS THE MOON EVER DONE FOR US? WOULD WE MISS IT IF IT WERE GONE? WHAT DIFFERENCE WOULD IT HAVE MADE TO LIFE ON EARTH IF THERE HAD NEVER BEEN A MOON?

These are the least of his problems. He finds that the sea has retreated many miles from the coast, since more of the Earth's oceans are gathered at the equator. A howling wind is shrieking across the land. A vast storm lashes the landscape with torrential rain and continuous lightning. Huge volcanoes erupt out of the crust, which is pierced by towering mountain ranges and abyssal canyons, and the ground is shaken and broken by many earthquakes. There is no sign of life, let alone human civilisation. There are no animals or plants, not a hint of greenery. Also, there is hardly any oxygen, so Al immediately starts to suffocate. He decides that he rather misses the Moon.

What has the Moon ever done for us?

Apart from helping to stabilise the tilt of the Earth so that it does not constantly change under the gravitational influence of Jupiter – playing havoc with our seasons – perhaps the most important effect of the Moon on the history of the Earth is that it has dramatically slowed the planet's orbit (see below). With only the Sun's tidal influence operating over the last 4 billion years, the Earth would be rotating at between 7 and 12 hours a day. The centrifugal forces would cause the planet to bulge at the equator much more than it already does, with dramatic tectonic effects. Extreme Coriolis effects (see p. 143) would pump energy into the atmosphere, causing high winds and storms. Life would probably never have evolved, and without photosynthesising bacteria and plants there would be very little free oxygen in the atmosphere.

Slowing the spin

The tidal effect of lunar gravity has slowed the Earth's orbit through friction. Imagine tying a balloon on a string to the belt of someone spinning on a revolving chair. The balloon would drag on the belt, and the belt would rub against the person – this friction would soak up some of their rotational energy, slowing them down.

069 THE DARK SIDE

There was something strange about the handsome young man, but Torhild couldn't put her finger on it. She had been milking the cows when she suddenly became aware of someone staring at her. Looking up, she saw the young man in his jaunty green and brown outfit leaning against the barn door, looking quite at home. He smiled at her, and asked a question, but she just blushed, and looked down.

Picking up the pail of fresh milk, she got to her feet and made for the door. The young man watched her all the way, turning to face her as she passed him. Up close, she could see his pale skin and freckles, and how his green cap was pulled down over his ears. There was something menacing about his smile, though he acted friendly enough.

In the yard Torhild turned to look at him. He kept back, out of the sun, facing her square on. A suspicion grew in her mind. She backed off slowly, and the lad moved forward. She stopped and he circled slowly around her, always facing square on to her. 'What's that you've got behind your back there?' Torhild called out to him. The lad just smiled. 'Why, pretty lady, I've absolutely nothing behind my back.' Torhild nodded. 'I know it well, just as I know you, for you *huldre* folk are hollow, with a space where your souls should be.' She crossed herself and the lad snarled, then in a blink he was gone.

'Everyone is a moon, and has a dark side which he never shows to anybody.'
– Mark Twain (1835–1910)

Synchronous orbit

Some of the varieties of *huldre*, or fairy folk, who haunted Orkney, the Faroe Isles and other Scandinavian settlements, were notorious for being hollow, with mask-like fronts and no backs or insides. To keep mortals from spotting this giveaway, they would have to turn always to face them, keeping their hollow backs hidden. The Moon plays a similar trick, because it is in synchronous orbit around the Earth. Its rotation exactly matches its orbit, so that as it goes around the Earth it turns, ensuring that the same side is always facing towards Earth, and the other always facing away. This means that whenever we see a new (i.e. no) moon, the far side of the Moon is bathed in sunlight. So it is a misnomer to talk of the dark side of the Moon: There is a hidden side, but it is no darker than any other part of the Moon.

070 INTO DARKNESS

Marcin suffers from skiaphobia, a pathological fear of shadows. All his life he has sought to avoid dark places, dusky corners and shady nooks. He insists on a battery of bright lights in every room in his house. He hates full moons, but strangely has never cared about being out in the middle of the night during a new moon. When it's completely black outside, he says, there are no shadows.

One evening he is hurrying to get home before the sun sets, when he notices something ominous on the eastern horizon. The sky is dark blue above him and a hazy pink above the eastern horizon, but on the horizon itself is a thin band of darkness. It seems to be getting bigger. Chilled to the marrow, he rushes home and turns on all the lights.

Later, Marcin ponders the disturbing phenomenon. Was it an approaching storm? A band of distant cloud? Turning on the news, he hears about an imminent lunar eclipse. The graphics make his blood run cold: They show a cone of dark shadow stretching away from the Earth towards the Moon, which is set to pass through the darkness that very night. The more he thinks about the Earth and its shadow, the more scared Marcin becomes. He begins to think he has a good idea of the nature of the dark band he saw, and he vows never again to go out at night.

Moon in shadow

Marcin has somewhat belatedly realised that the biggest shadow cast on Earth is that of the Earth itself. Nighttime is nothing less than the shadow of the Earth, cast over the side of the planet facing away from the Sun. The huge bulk of our planet blocks the sun from a cone of space that extends far into space: The Earth's shadow extends for about 1,400,000km (870,000 miles). When the Moon's orbit takes it directly through this cone (which can only happen during a full moon), this is called a lunar eclipse.

Anti-twilight

To get a more direct appreciation of the Earth's shadow, look for a common but little observed phenomenon. As the sun sets or rises, falling or climbing below the horizon, the curve of the Earth casts a shadow onto the atmosphere on the opposite horizon. Given a clear view of a wide horizon, ideally with a long sight line from a high vantage point (an aeroplane is best), the Earth's shadow can be seen as a very dark blue band just above the horizon, stretching all the way across it. Above it may be a band of pinkish colour, known as the anti-twilight arch or Belt of Venus, caused by reddened sunset/dawn light scattering off dust particles.

'These late eclipses in the sun and moon portend no good to us.'

– Shakespeare, *King Lear* (c. 1605)

071 BLACK HOLE

Astronauts Wu, Yusef and Zulaka live on a space station in deep space. Due to an unfortunate series of coincidences, they find that the station has drifted between two black holes: one small and one supermassive. Wu and Yusef start to panic. Of course, they can't actually see the black holes themselves: The gravity of each has warped the fabric of space–time into a kind of hole – a singularity – from which light itself cannot escape. But not only can they see dark shadows blocking out starlight, they can also see the distorting effects of the presence of the black hole, with light from stars and galaxies behind them bent around the holes as though they were giant lenses. Even more disconcerting, they can see the event horizons of the black holes: The threshold at which the gravitation becomes so strong that light cannot escape. Quantum effects mean that as matter falls into the black hole, streams of particles are radiated back out from the event horizon – this is called Hawking radiation – so the event horizon appears to be glowing.

In their panic, Wu and Yusef decide that the space station is doomed, and that they have to get out of it as soon as possible. They put on their spacesuits and prepare to fling themselves out of the airlock. 'Stop!' Wu says to Yusef. 'You're jumping towards the supermassive black hole.' Yusef replies, 'Rather that than jumping towards the small one.' They leap from opposite airlocks and fall into the gravitational clutches of their respective holes. Zulaka watches as they spiral closer to their respective fates. What will happen to them?

WHAT WOULD HAPPEN IF YOU FELL INTO A BLACK HOLE?

Spaghetti time

Wu, falling into the small black hole, does not last long. With a small black hole the difference in gravity experienced by his head and his feet is tremendous, with the result that he is stretched like a piece of spaghetti and torn apart before he even reaches the event horizon.

Death on the event horizon

Zulaka, watching Yusef from the space station, sees him warp and stretch as he gets closer to the black hole, but she also sees him slow down as he approaches the event horizon. As he reaches it, apparently smeared out into a thin layer, he freezes. To Zulaka, the heat of the Hawking radiation means that, over time, Yusef will appear to evaporate into nothing.

The other side

Bizarrely, Yusef experiences things very differently. The event horizon is not a physical barrier – it is something that only outside observers see. Yusef himself will not notice it, and will simply continue in free fall until he reaches the singularity in the centre of the black hole. If the black hole is big enough, this could take a lifetime. For Yusef, time and space seem to proceed as usual, and he will survive until his air runs out, falling freely through space.

072 BLAST-OFF

'What's the point in having it, if we never get to use it?' asks Kerechenko. He is standing on the eastern rim of Engel'gardt crater. 'And this is the best spot for it. They call this the Selenean summit; it's the highest point on the surface of the Moon. I've got a clear line of sight for aiming!' Kerechenko is waving a three-barrelled pistol with a folding stock, standard issue to Russian cosmonauts.

'I just think it's a bad idea,' protests Yarmukov. 'We don't know what will happen – we don't even know if it will go off.' Kerechenko dismisses his concerns: 'It'll fire alright. Gunpowder contains its own oxidant, so no atmosphere is needed. Plus I've worked out the physics – and they're pretty cool.' He explains that, with no air resistance, the bullet could potentially speed on forever, without ever slowing down.

'But you're forgetting Newton's third law,' argues Yarmukov. 'When that bullet blasts out of the barrel, you will shoot off in the opposite direction.'

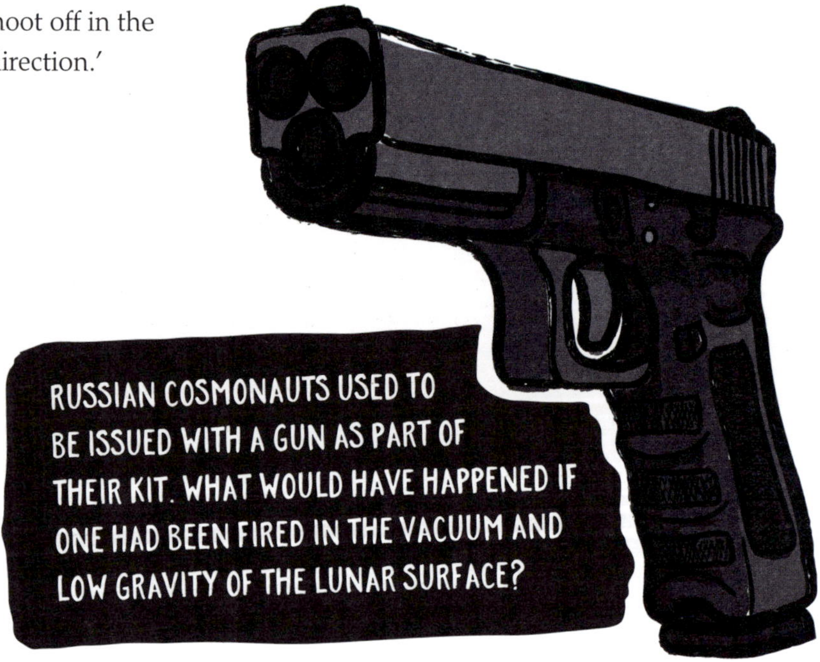

Kerechenko indicates that he is well braced against a low boulder. 'The bullet may be travelling at almost 1,200m/s (4,000 feet per second), but its mass is tiny – the same force working on me will be easy to handle.'

Yarmukov is not convinced, but Kerechenko ignores him, levelling his gun at the distant horizon and pulling the trigger. He rocks backwards as a sphere of smoke expands from the tip of the barrel. The bullet races away over the horizon. 'Ha ha!' crows Kerechenko. 'How cool is that? It probably won't stop until it reaches Mars!'

But Yarmukov has been doing some figuring out. 'You forgot about the Moon's own gravity. It may be weak, but still, according to my calculations the escape velocity for an object to leave lunar gravity is more than 1,200m/s (4,000 feet per second). And if it doesn't have enough speed to leave lunar orbit, it will go all the way around the Moon and come ba— ' Two holes appear in Kerechenko's helmet, one at the back and one at the front. As his spacesuit depressurises, blood and grey matter spray out through the hole in his visor, freezing immediately into tiny globules of ice.

Yarmukov switches on his suit radio. 'Mission Control, we have a problem . . .'

Newton's cannon

In 1687, Isaac Newton published in his book *Principia* an illustration of a cannon, firing horizontally from the top of a mountain. He explained that Earth's gravity will make the cannonball follow a parabolic path down towards the ground, but that with enough launch power it will travel so fast that it will leave Earth's orbit before this can happen. Setting aside air resistance, if it is fired at precisely the right speed – about 26,000km/h (16,000mph) – it will fall towards the Earth at just the same rate at which the surface of the planet curves away, constantly missing the Earth. In the Moon's weaker gravity, a projectile needs to be travelling at about 1,200m/s (4,000 feet per second) to go all the way around, as Kerechenko has found to his cost.

073 VACUUM CHICKEN

The year is 2162. The human race has developed advanced technologies and explored the Solar System, but teenagers still do stupid and reckless things for laughs. Bojohn and his friends are mucking around with a teleporter, playing a game called Vacuum Chicken. The aim of the game is to pick the most exotic place in the Solar System to which to teleport – without a space suit. The teleportee has to last as long as possible before passing out, which triggers the automatic recall to transport them back.

First up is Trijean, who doesn't trust the teleporter's range reliability and opts for a timid low Earth orbit. In 10 seconds she reappears, unconscious, with blood gushing from her mouth and nose, her face and hands swollen and rimmed with frost. 'Oooh! Epic fail,' groan her friends, loading her into the organic repair chamber. 'I reckon she tried to hold her breath,' comments Vishwu. 'Schoolgirl error.' He steps up and programmes in a high tropospheric altitude on Venus, disappears through the teleportal and reappears 30 seconds later, his skin hideously blistered and peeling. As they load him into the repair chamber, Vishwu recovers consciousness enough to gasp, 'It was going well until I fell through a cloud of sulphuric acid.'

WHAT WOULD HAPPEN IF A SPACEWALKING ASTRONAUT TOOK OFF HIS OR HER HELMET? WOULD THEY SUFFER LIKE ARNOLD SCHWARZENEGGER AT THE END OF *TOTAL RECALL*?

Next to go is Palwar, who chooses the surface of Titan, one of the moons of Saturn. A minute later she reappears, her skin blue and hard, a cloud of frozen exhalation still stuck to her lips. Finally, Bojohn approaches the teleporter. 'Got to get this just right,' he mutters, as he inputs into the controller extremely precise coordinates for the interior of the Jovian moon Ganymede. He steps into the teleporter and disappears. Three long minutes pass before he rematerialises, soaking wet and shivering uncontrollably. 'I couldn't take it any longer,' he says through chattering teeth, 'but I think that's a new record.'

The air up there

Contrary to Hollywood standard practice, losing your helmet in space will not cause your head to explode. In a vacuum, the pressurised air in your lungs will try to force its way out; if you try to hold your breath you could cause serious damage. Once the air is expelled, you have about 15 to 45 seconds (opinions differ) before you use up the oxygen stored in your bloodstream and pass out. You will suffocate a couple of minutes later. Your skin will keep the rest of you intact and, though exposed portions will freeze, it takes a long time for heat to radiate from your body in a vacuum. You may well swell up, your saliva will boil away, and you will probably feel a fizzing sensation from this and from bubbles of gas forming in your body fluids, which could cause a case of 'the bends'.

Where in the Solar System?

There are a few places in the Solar System where the pressure is sufficient to allow someone to keep the air in their lungs and survive as long as they can hold their breath. About 52.5km (32.6 miles) up in the atmosphere of Venus, the air pressure is about 65 percent of that of Earth, although clouds of sulphuric acid make it uncongenial. On the surface of Titan the pressure is about 1.5 times that on Earth, but the temperature is a chilly –179°C (–290°F). Beneath its icy shell, Ganymede has liquid-water oceans in which it might be possible to hold one's breath.

074 STAR CRUISING

Anander, Cho and Mara (see p. 146) are students on an astronomy course. They have been set a new assignment: to come up with novel ideas about how to cross the apparently unbridgeable gap to the stars. The nearest star to our Solar System is Proxima Centauri, about 4.22 light years from Earth. At its top speed, the fastest man-made object ever, the Helios 2 solar explorer, would take 19,000 years to reach Proxima Centauri. The students are asked to think about interstellar craft designs that could shorten the journey time.

Anander suggests a Bussard ramjet. This is a spaceship with an enormous magnetic scoop projected in front of it, which collects hydrogen gas from interplanetary space in order to burn it in a fusion engine. He also considers an Enzmann starship, which carries a 2.7 million metric-ton (3 million-ton) sphere of frozen hydrogen on its nose and, again, uses it to fuel fusion energy.

Cho suggests a nuclear pulse engine. This essentially explodes a series of hydrogen bombs behind it to blast it forward, accelerating to great speeds. Or, for even more explosive sources of energy, Cho theorises that a rocket could bring together matter and antimatter, and use the resulting fantastically energetic annihilation to power a rocket.

Mara thinks the answer is to avoid travelling across space in the conventional sense. She suggests that a ship that could warp the fabric of space–time ahead and behind itself could travel in a 'warp bubble'; rather than moving the ship through space, the ship remains stationary and space is shifted instead.

ARE THERE TECHNOLOGIES THAT COULD ENABLE US TO CROSS INTERSTELLAR SPACE?

Need for speed

In order to reach speeds that make interstellar travel feasible, the mass of a starship must be accelerated. This requires fuel of some sort, and this requirement is one of the primary limiting factors in starship design. Anander's Bussard ramjet was initially proposed as a solution to this problem, but it is possible that it would create more drag with its scoop than the thrust it could generate. The Enzmann starship faces the problem of inertia – accelerating the vast mass of hydrogen would take a long time. By the time the fuel had been burned through, the craft would be much lighter and could be travelling at colossal velocity, but the trip would be a long one.

Running on empty

Even high-efficiency rocket engines, such as the nuclear pulse engine or the antimatter engine, would burn through immense amounts of fuel. Assuming that the starship is accelerated to a speed high enough to reach Proxima Centauri in 900 years, a nuclear pulse engine would require 1,000 supertankers of fuel, and an antimatter engine would need 10 railway cars' worth of fuel. At the moment only tiny, subatomic particles of antimatter can be generated, and storing it would be a considerable problem. Warp drives, such as the one proposed by Mara (known as an Alcubierre warp-bubble drive) are purely speculative, and would probably require astronomical energies.

Pipped at the post

A spaceship that sets off on a very long voyage could well find itself overtaken by a technology developed many years later.

THE NATURAL WORLD

'It seems to me that the natural world is the greatest source of excitement; the greatest source of visual beauty; the greatest source of intellectual interest. It is the greatest source of so much in life that makes life worth living.'

– Sir David Attenborough (b. 1926)

075 COME INTO MY PARLOUR

It is the year 2060. Extensive nuclear testing in the Australian outback has had a number of unfortunate side effects, the most headline-grabbing of which has been the giant mutant bugs marauding through cities. Fortunately, most of the monsters have been dealt with, but some remain on the loose. Captain Selly is unconcerned, however, despite the fact that the jumbo jet he is piloting is about to touch down at Woollomooloo Airport, just beyond the Condemned Zone.

'It'll take more than a big bug to stop this beauty,' he boasts to his copilot. 'She weighs more than 200 metric tons (220 tons), with an airspeed of around 200km/h (125mph) even when she slows down to land. Now I'm no Einstein, but I reckon that gives her a momentum of over 11 million kg m/s (80 million foot-pounds per second). Where's the creepy-crawly that could get in our way without being splatted?'

At that precise moment the copilot notices something weird in the distance. 'Captain, doesn't that look like . . . well, I mean it can't be, of course, but doesn't it look a bit like a giant web?' By the time Captain Selly has finished scoffing, the jet has plowed into the colossal web. 'Don't worry!' he reassures his copilot. 'There's no way something that comes out of an animal's butt can possibly take this much strain—'

But his voice dies in his throat as he and everything else in the plane are flung forward with great violence. The silk gives and gives until it seems it must break, but despite stretching for several miles it holds. Soon the jumbo jet is stuck fast in the web like an enormous fly. Captain Selly and his copilot start to wonder what kind of creature could have made a web this big . . .

Draglines

Spiders produce different types of silk for different purposes, and their properties vary with water content and species. The strongest type is 'dragline' silk, which spiders use for safety lines, and which has a tensile strength (the amount of stretching it can take before breaking) greater than most types of steel. Arachnophile Ed Nieuwenhuys has calculated that a dragline 30km (19 miles) long would be strong enough to stop a jumbo jet, despite being only as thick as a pencil. He also calculates that it would require 102 billion garden spiders to spin a web this big, but with the assistance of nuclear testing-induced mutation, perhaps a single giant arachnid might suffice?

SPIDER SILK IS RENOWNED FOR BEING STRONGER THAN STEEL; ITS TENSILE STRENGTH MATCHES THAT OF PARA-ARAMID SYNTHETIC FIBRES USED TO MAKE BODY ARMOUR. SO, JUST HOW STRONG IS IT?

076 MULTIPLE CHOICE

Oksana has been assigned by the Central Ministry for Total Employment to work at the Central Library in the Central District. She joins the queue of other drones reporting for their make-work drudgery. When she reaches the head of the queue, the Vice-Deputy Assistant Subsection Underlibrarian for Dewey Numbers 113.84.456 to 113.84.457 hands her a pile of books and a sheet of instructions, and then points to a stretch of empty shelving in the distance. Oksana shuffles over, puts down her books and reads the piece of paper.

'Welcome to your exciting new career opportunity. You have been given 15 different books. Your task will be to arrange the books in a line on the shelf. Then rearrange them in a different order. Keep doing this. The only rule is that you must never arrange them in the same order twice. Rest assured that your new career offers a lifetime of worthwhile and fulfilling employment.'

She looks at the 15 books in disbelief, and then whispers to the elderly man at the next shelf, who is patiently rearranging his own 15 books, 'This is ridiculous. How long is this supposed to keep me busy if I'm never allowed to repeat the same order?' The old man finishes arranging his books, then picks up 14 of them and shifts them around. 'I've been doing this job for the last 67 years,' he croaks, 'and I have another 12½ million years to go. You'd best get started.'

IT IS OFTEN SAID THAT NO TWO SNOWFLAKES ARE THE SAME. IS THIS TRUE, AND IF SO, WHY?

Points of difference

Oksana has failed to do the mathematics entailed by her new task. With 15 books, there are 15 choices for the first book in line, 14 for the second, 13 for the third, etc. So the total number of possible arrangements is 15 x 14 x 13 . . . etc., known as 15 factorial (written as 15!). This works out to about 1.3 trillion arrangements. When constructing a snowflake, nature has many more options for different arrangements than this, because each snowflake has dozens if not hundreds of points of potential difference. This means the total number of possible configurations of snowflake is far greater than the number of atoms in the universe, so that while it is theoretically possible that the same pattern may have appeared twice, it is astronomically unlikely – even though an estimated 10^{34} snowflakes (10 billion trillion trillion) have fallen in Earth's history.

Levels of complexity

There are, however, different 'levels' of snowflakes, more properly known as snow crystals. While the classic branching crystals, known as dendrites, do indeed come in a practically infinite complexity of forms, there are small, simple snow crystals that start off as just hexagonal prisms or plates, and these do look alike.

'. . . the endless repetition of an ordinary miracle.'

– Orhan Pamuk, *Snow* (2004)

077 LONG SHOT

Chuck has just bought a lottery ticket. He steps out the door of the supermarket clutching his ticket. He has a good feeling about this. Just this once, he feels like he might actually be in with a shot of winning the jackpot.

Crossing the road, he is hit by a car and knocked into a nearby creek, where he is attacked by an alligator. He stumbles out of the creek but suffers a lethal sting from a hornet. As his throat begins to constrict, his phone rings. He answers it to be told that he has just been selected to play for an NBA team. He hangs up and attempts to dial the emergency services, but in his parlous state is only able to bash the keys randomly. He gets through to Miss Universe.

A meteorite lands on his foot. Then a chunk of aeroplane lands on his other foot. His hand is digested by flesh-eating bacteria. A messenger arrives bearing news that he has won an Oscar, and another brings a note from NASA to say that he has been selected for the astronaut programme. He wins a gold medal at the Olympics and is elected president of the United States.

Staggering onto a golf course, Chuck accidentally knocks in a hole in one, and then falls into the sea where he is attacked by a shark. Crawling back on land, he tries to use a product designed for right-handed people, despite being left handed, and sustains another fatal injury. Finally he is struck by lightning, and crushed by a falling vending machine.

'Adventure upon all the tickets in the lottery, and you lose for certain; and the greater the number of your tickets the nearer your approach to this certainty.'

– Adam Smith (1723–90)

Against the odds

Obviously all lotteries are different, but, for example, the odds against winning the jackpot in the Powerball lottery in the United States are 1 in 175,223,510, and for the National Lottery jackpot in the UK, 1 in 45 million. Each of the things that happened to Chuck is more probable than this. Even with the more favourable odds offered by the Irish Lotto jackpot (one in 10.7 million), you are still up to 8 times more likely to be hit by lightning.

'My wife said to me: "If you won the lottery, would you still love me?" I said: "Of course I would. I'd miss you, but I'd still love you."'

– Frank Carson (1926–2012)

ARE YOU REALLY MORE LIKELY TO GET STRUCK BY LIGHTNING THAN WIN THE LOTTERY?

078 STRANGE ATTRACTION

Danno's Dodgems is a veteran bumper-car concession at the fairground. Danno wants to increase his earnings, but he can only fit so many bumper cars onto the rink at any one time because of the violence of their collisions. They ricochet off each other with so much energy that if he packs in too many cars they start flying off the rink.

Danno has a cunning plan. He puts magnets on all four corners of each car. The two magnets on the front of each car have opposite polarity to the ones on the back. Now, as the cars come near one another, they tend to become magnetically stuck to one another. Danno's mate Benno objects that this will simply ruin the bumper-car experience, but Danno points out that his rink is powered by a very high-powered battery, so that the cars move about so fast that the relatively weak attraction between the magnets is easily broken. Cars will tend to draw closer together than they would otherwise, increasing the density of cars he can achieve on the rink, but

without really interrupting the bumping fun. Sure enough, Danno's clever trick works a treat and he is able to pack in 15 percent more cars. Everyone agrees he is a genius.

But then something unexpected happens. With so many cars on the rink, Danno's battery runs flat. When the cars are hardly moving at all, each magnet on each corner of each car sticks to another magnet on one of the other cars. Each car thus sticks to four other cars, but the only way this can happen is for the cars to arrange themselves in a lattice in which the space between the cars is much greater than it was before. Instead of squeezing in closer together, the bumper cars suddenly spread out to a much lower density and the ones on the outside all fall off the side of the rink.

Why ice floats

Danno's bumper cars are analogous to water molecules, which carry the equivalent of four tiny magnets. In a liquid, when the molecules have quite a lot of energy and are whizzing about and bashing into one another, these 'magnets' attract one another, forming, breaking and re-forming weak bonds at a dizzying pace. This makes water molecules much stickier than other similar molecules, and means that liquid water is denser than would otherwise be expected. When the temperature drops and the molecules slow right down, each molecule forms bonds with all four of its 'magnets' at once, and to accommodate this the molecules spread out into a lattice crystal formation with much more space between the molecules. This is why ice is less dense than water, and thus floats instead of sinks.

Ice blanket

Because ice floats, when the top of a body of water freezes the ice forms an insulating layer preventing the rest of the body from freezing. This is one reason why only a fraction of the ocean freezes, even in the polar regions. Salt, currents and geothermal energy also limit ocean freezing.

079 NEED FOR SPEED

Plippy and her friends gathered to watch the big drops going past. 'Wow, did you see that one?' 'Look how fast they're going!' Plippy said dreamily, 'I wish I could fall that fast.' Her friends teased her: 'How's a little drip like you ever going to catch up with the big ones?' 'Yeah, you're way too small, Plippy; you're only drizzle!'

But Plippy was stubborn and determined; she *would* go as fast the big drops when it came to her turn to fall. Her mother sighed, 'Ever since you were no more than a seed, I've tried to explain to you the way it works. You simply can't fall faster than your size will allow. I know it's not fair, but it's physics!'

Finally the big day came. The cloud had grown too big and got too high: It was time for the big rain. Plippy watched her little friends drop away one by one, each floating down at a gentle speed, buffeted by air currents and passing thermals, hardly seeming to fall at all. But she refused to give up on her dream. She waited until a really big drop plummeted past and then she leaped, falling in close behind the larger blob of water. It seemed to work! She was getting faster, staying close behind the big drop, so close they almost merged. She sped past her friends, squealing with glee, and still she accelerated. Faster, and faster still, until the whole world was just a blur of speed. Finally she pushed clear of the big drop and felt the sudden, violent resistance of the air, trying to slow her down. But it was too late, she had done it, she was going faster than any small drop had gone before, and she was going to make one hell of a splash.

HOW FAST DOES A RAINDROP FALL?

Terminal velocity

The speed limit for a raindrop is normally determined by its terminal velocity – the speed at which the air resistance it encounters (which increases with speed) matches the force of gravity (which does not increase with speed), so that the drop no longer accelerates and cannot fall any faster. Terminal velocity is greater for large drops than small ones. Large drops around 5mm (0.2 inches) across fall at a maximum speed of about 9m/s (20mph), while small, drizzly drops around one-tenth of that diameter should be limited to about 1m/s (2mph).

Super-terminal

Plippy has worked out a trick that has only recently been discovered by raindrop researchers. Laser measurement of rainfall has shown that 30 to 60 percent of small raindrops are 'superterminal', meaning that they fall faster than their predicted terminal velocity. Researchers believe that the small drops are fragments of larger drops that split off at the last minute and haven't yet slowed down to their terminal velocity limit, or which fall in the wake of bigger drops, taking advantage of reduced air resistance. Some of the drops travel 10 times faster than their theoretical limit.

'Some people walk in the rain, others just get wet.'

– Roger Miller (1936–92)

080 HEAVY WATER

The people of Far Far Away were being terrorised by a fierce and colossal ogre. They sent for Jack the Giant Slayer, and told him of the gruesome giant. 'He's as big as a castle,' they said, 'and as strong as a hurricane. He has the appetite of a whale and the thirst of a desert. He's as cruel as winter and as tough as old leather.' 'Has he no weakness?' asked Jack. 'He is vain and boastful,' they said.

And so Jack arrayed himself in finery and went forth to the giant's lair, attended by a choir of heralds in livery. He had them sound silver trumpets within the giant's earshot and proclaim that 'Jack the Giant Slayer' was passing by. The ogre rushed out and brandished his club. 'Giant Slayer, is it? You have never met my like.' But Jack turned his back on him, saying, 'You do not seem worth fighting, for now that I see you I take pity, so scrawny and weak are you.'

This enraged the giant, who danced about plucking up trees and smashing boulders to dust. 'I am the strongest ever to walk this land,' he boasted. 'Set me a test and I will show you.'

CLOUDS APPEAR TO BE WEIGHTLESS, FLOATING THROUGH THE INSUBSTANTIAL UPPER AIR, BUT DO APPEARANCES DECEIVE? JUST HOW MUCH DOES A CLOUD WEIGH?

Jack puffed out his cheeks disparagingly: 'Why, I'll wager you could not even hold up as much weight as a cloud, so puny are you.' The giant shrieked: 'What?! Those balls of candy floss that do hang and float in the air? I could hold up a thousand.'

'Very well,' said Jack. 'Prove it.' He led the giant to an enormous great rock the size of a hill, and bade the giant lift it. The ogre protested, but Jack explained that this boulder was of the same weight as a single cloud. Unless the monster wanted to be proved a liar as well as a weakling, he must fulfill his boast. The ogre could not bear to be called either of these things, and so he attempted to lift the vast rock and was squashed flat in an instant.

What's in a cloud?

The giant has fallen for a common misconception: that because clouds float in the atmosphere, they must be lighter than feathers. Floating, however, relates not to weight but to density. The moist air of a cloud is marginally less dense than the dryer air around it. A cloud, for instance, might have a density of 1.134kg/m^3 (0.0708 pounds per cubic foot), while dry air has 1.14kg/m^3 (0.0712 pounds per cubic foot). But because a cloud is such a vast pocket of air, its total weight is colossal. A cumulus cloud might be more than a kilometer (half a mile) on each side, giving it a volume of 1 cubic kilometer (0.24 cubic miles). At the density given above, such a cloud would weigh more than 1 million tons.

Water content

Another way to define the weight of a cloud is in terms of its water content – this is, after all, what distinguishes it from the surrounding air. The water density of the cumulus cloud discussed above might be around 0.5g/m^3 (0.00003 pounds per cubic foot), giving a total water content of 453.6 metric tons (500 tons), not far off the weight of a fully loaded jumbo jet.

081 SAFE AS HOUSES

The Bilateral Organisation for Operations and Maintenance is inviting tenders for a lucrative contract. International bidders may pitch for the chance to be the storage site for the world's combined high-level nuclear waste repository.

The first pitch comes from Fiji, which wants to import the world's waste and dump it on an isolated island in the middle of the Pacific. Italy offers to store it in a deep mine shaft. Australia touts its stable government, developed infrastructure and highly suitable geology. A site in New Mexico in the United States also claims to have suitable geology, and points out that it already possesses a well-established underground storage facility.

CURRENTLY, MOST OF THE WORLD'S HIGH-LEVEL NUCLEAR WASTE IS KEPT WHERE IT IS GENERATED, AT NUCLEAR POWER STATIONS. IN THE LONG TERM, HOWEVER, WHERE SHOULD RADIOACTIVE WASTE BE STORED?

A political problem

The technology for long-term storage of radioactive waste with a high degree of security mostly exists already – in the United States, for instance, there is indeed an underground storage site near Carlsbad, New Mexico, which ticks almost all of the boxes for long-term disposal. Another site identified as viable for storage of all the waste in the world is in South Australia. The real obstacles to identifying long-term storage solutions are political.

Best of a bad bunch

In the popular conception, radioactive waste is the worst and most pernicious by-product of humanity's technologically advanced civilisation. In fact, a case could be made that, relatively speaking, it is not so very bad. OECD countries produce around 300 million metric tons (330 million tons) of toxic waste per year, of which just 12,000 metric tons (13,000 tons) is high-level radioactive waste. Unlike most of the rest of the toxic waste, which remains toxic indefinitely, radioactive waste gets less dangerous over time. Precisely because it is radioactive, the waste is always transmuting into less radioactive elements. For instance, strontium-90 and cesium-137 have half-lives of about 30 years, so that in 30 years waste contaminated with these isotopes will only be half as radioactive.

Winning bid

Some of the bids the WAA receives are ill-advised. Fiji may have the advantage of being isolated in the middle of the Pacific (one serious proposal for disposing of high-level waste is to sink it in a deep Pacific Ocean trench where it will be subducted into the Earth's mantle along with the oceanic crust), but shipping waste across storm-tossed oceans and dumping it in porous rock formations with a high water table is rash. Deep boreholes are considered to be good options for permanent disposal, but they would have to be in areas offering geological and tectonic stability over thousands of years. Italy is in a tectonically active part of the world, whereas Australia has tectonically inactive geology and a good supply of deep-lying strata with a low degree of groundwater mobility.

082 THE AIR UP THERE

Irréflèche Montgolfier, the little-known cousin of the celebrated Montgolfier brothers, has devised a balloon design of his own. His balloon can hold far more people than the paltry basket of the *globe aérostatique* of Joseph-Michel and Jacques-Étienne, thanks to his revolutionary expanding wicker technology. Irréflèche has developed a new form of wicker, which, when heated, expands without losing strength, enabling him to build a truly grand vessel to be slung beneath the envelope of hot air that will bear it into the sky.

On the day of the launch, over a hundred brave and curious souls pack into the wondrous vehicle, whereupon Irréflèche signals to the groundsmen to release the tethers, and his magnificent device soars skywards.

At first, everyone is enchanted by the rapid ascent, but Irréflèche grows uneasy as the breath of his passengers starts to become visible as clouds of vapour. There is an ominous creaking sound from the gondola beneath their feet, and Irréflèche stoops to observe it closely. With his unaided eye he can see the fibres of wicker contracting as they cool.

WAIALEALE IN HAWAII EXPERIENCES AN AVERAGE OF 335 RAINY DAYS A YEAR, WHILE ARICA IN CHILE HAS JUST ONE DAY OF RAIN EVERY 6 YEARS. INHABITANTS OF BOTH PLACES MAY WELL ASK THEMSELVES, WHY DOES IT RAIN?

Moments later there comes a cry as passengers nearest the edge of the giant basket are knocked backwards by the rapidly constricting rim. The cries turn to screams of terror as the already crowded mass of people is crushed together, followed by appalled exclamations as those no longer able to fit within the shrinking gondola begin to fall from its sides. Down below, horrified onlookers rush for cover as it begins to rain people.

Higher is colder

Irréflèche's poorly conceived gondola is analogous to a pocket of air. Just as his gondola can hold more people when it is warm, so air can hold more gaseous water when it is warm. Like the balloon, warm air is less dense than cooler air and so will rise. But because the density of air diminishes with altitude (and with it the ability to hold heat energy), and because air is mainly heated from the ground, the atmosphere cools with altitude – 6.5°C (11.7°F) for every 1,000m (3,280 feet). So, as the air rises it cools and its capacity to hold water vapour diminishes, as does the capacity of the gondola to hold passengers. Above a certain height the vapour in the pocket of air will condense into water droplets, forming clouds, and when the droplets become heavy enough, the downward force they exert under gravity overcomes the upwards force of air draughts and air resistance, and they fall as rain.

Types of rain

- Convective rain results from pockets of warm, wet air rising because they are less dense than the surrounding cooler air.
- Orographic rain results from warm, wet air that is forced upward by passage over mountains.
- Frontal rain results when a mass of warm air runs into a mass of cold air. The warm and cold air masses don't mix; instead, the warmer one is forced up and over the denser cold air, creating a weather front, which gives frontal rain.

083 BEAR NAKED

Polo Chanel, the most daring polar-bear fashion designer of the age, peeked out at the expectant audience sitting around the bearwalk and smiled to herself. She would give them a show to remember. The voice of the announcer came over the PA as the models began to appear.

'Showcasing our most classic look, Perdita is wearing dazzling white thanks to her outer coat of hollowfibre hair. Designed by nature, these outer hairs are perfect at scattering incident sunlight in every colour of the spectrum. And of course, when you combine all the colours you get white! Perdita wears her newest coat for the start of the season, and is best viewed in direct sunlight for a genuinely luminous colour experience.

'Next on the bearwalk, Paria models a yellow coat for the more mature customer. Paria's outer hairs have been carefully stained with a season's worth of seal oils derived from her rich diet. And following her is Patricia, modelling a shocking new look, incorporating algae into the hollow shafts of her guard hairs to produce a striking green coat. It takes long months of exposure to a warm, humid climate to achieve this look, ladies and gentlebears.

ANY CHILD WILL CLAIM TO KNOW THE ANSWER TO THE QUESTION, 'WHAT COLOUR IS A POLAR BEAR?' BUT JUST BECAUSE THEY APPEAR WHITE, DOES THIS REFLECT THEIR TRUE SKIN COLOUR?

'And now we come to our finale . . .' The crowd hushed in expectation, then burst into gasps of mingled surprise and appreciation, as an entirely shaven, black-skinned bear appeared. 'Prudence has been close-shaved from head to claw to achieve a completely naked look.' The crowd rose as one to applaud Polo's bold gambit, although they all agreed that the average bear on the ice floe would never really go for it.

Hollow hairs

Polar bears are brilliantly adapted to their snowy environment, at least as far as the visible spectrum is concerned. Light is an electromagnetic wave (an oscillation in an electromagnetic field). More specifically, 'light' is the name we give to the visible part of the range – or spectrum – of different lengths and frequencies of electromagnetic wave. The spectrum spreads all the way from waves with very low frequencies and long wavelengths, such as radio waves, to those with very high frequencies and short wavelengths such as X-rays and gamma rays. Just beyond the fringes of the visible spectrum are infrared and ultraviolet light. Some of the polar bear's prey animals, such as reindeer, can see into the ultraviolet part of the spectrum, where polar bears are much more visible. The transparent hollow guard hairs of their outer coats scatter light across all wavelengths, making them look white, and this effect is enhanced by colourless inner coats, light-scattering particles inside the guard hairs, and salt crystals from seawater between the hairs. Underneath their fur, however, polar bears have black skin.

Multicoloured bears

At their whitest when they have molted their old coats and grown in their new ones, polar bears can acquire other colours, including yellow from oils in their diet that are deposited in their hair, and green from algae acquired when they live in zoos in warmer climates.

084 GOOD VIBRATIONS

Being a temple guardian in the service of Bastet, the Egyptian cat goddess, is stressful and demanding work. Of course she doesn't speak our human language, so we have to do the best we can to interpret the noises she makes. Hisses and spitting are pretty straightforward, but the worst is the purring. Sometimes it can be relaxing, but at other times it can be shrill and distressing, putting your teeth on edge. At times she seems to be using it to talk to the cats that come and go freely in the temple; at others, Bastet seems to be trying to talk to us – or at least, to issue demands, usually for food.

There was one occasion when I was injured in a battle and broke some ribs. In great pain, I was carried into the throne room and laid on a dais in front of the goddess. She began to purr and I could feel the vibrations passing through my bones. The pain ebbed and I felt relaxed and comforted. I stayed there for several days, with Bastet purring all the while, and my injuries healed with remarkable rapidity. Since then I have enjoyed more than ever the purring of the cats who visit the temple, and I feel that perhaps I understand my goddess a little better when she purrs.

CATS PURR NOT ONLY WHEN CONTENT, BUT IN A RANGE OF SITUATIONS. PURRING TAKES ENERGY, AND SO IT MUST PERFORM SOME USEFUL FUNCTION OR, HAVE PROVIDED AN EVOLUTIONARY ADVANTAGE. WHY DO CATS PURR?

Comfort, food

For such a familiar phenomenon, cat purring is surprisingly poorly understood. There is still some debate about how cats purr (for instance, whether or not it involves the diaphragm), although it is known that they purr during both inhalation and exhalation. Theories concerning the purpose or utility of purring include communication and comforting, for cats are observed to purr at their kittens, who can purr back from only the second day of life, and they can nurse and purr at the same time. Cats purr both when contented and when stressed, such as when injured or when visiting the vet, so it may be that purring can be self-comforting. One suggestion is that purring is linked to release of endorphins in the cat's brain. Endorphins are naturally occurring opiate neurotransmitters that stimulate reward centres in the brain to create pleasurable sensations, and are also analgesic. This could explain why purring happens in response to both positive and negative stimuli. Researchers have also discovered that, to solicit food from their owners, cats use a different tone of purr than when being stroked.

Osteo-purr-osis

One of the most startling theories is that cat purrs help stimulate healing, particularly knitting of bones. There is an old veterinarian's saying that a purring cat can heal a room of broken bones, and cats themselves are renowned for rapid recovery from wounds and surgery. Domestic cats purr at a frequency of about 26 hertz, which has been linked to tissue regeneration. High-impact exercise is known to promote increases in bone density, and purring possibly works in a similar way, by creating pressure waves. It could be that purring is a form of acoustic vibrational therapy that cats employ in their downtime, to ensure that their bones are strengthened for when they are hunting.

085 SIGNS OF INTELLIGENCE

Three travellers to the newly discovered Planet X have returned accompanied by three different alien species: an ajetuink, a bufeget and a crinitex. All three claim to have taught their alien companions to speak sign language.

Professor Awaii demonstrates by pointing to some food and signing the word 'food' to the ajetuink. The ajetuink signs the phrase 'Give me,' and the professor gives it some food. Awaii then signs the phrase, 'How many people are in this room?' but the ajetuink looks at him blankly.

Professor Brindle signs the same question to the bufeget. The bufeget watches carefully and then starts to count by holding up its pseudopods. Brindle and the other nine people in the room watch intently. When the alien gets to nine they hold their breath to see if it will continue or stop, and when the bufeget holds up a tenth appendage they all break out into grins and start nodding their heads. The bufeget stops counting.

Professor Casiewicz signs 'Hello, how do you feel?' to the crinitex. The crinitex signs back the word 'tired'. Casiewicz shows it a red ball and a blue cube, which it has learned to name. Then she shows the crinitex a blue ball and a red pyramid, neither of which it has ever seen. The crinitex correctly describes the colour of the ball and when asked about the pyramid says, 'Not ball, not cube.' Asked to give it a name, the crinitex says 'pointing cube'. By now the crinitex has had enough, and signs 'Many people, crinitex tired, people go now'. The humans leave the room and watch the three aliens through a two-way mirror. They observe the crinitex signing to the other two aliens: 'This planet is the worst.'

CAN CHIMPANZEES REALLY USE SIGN LANGUAGE?

Clever Hans

The ajetuink and bufeget demonstrate some of the problems with interpreting claims about language abilities in non-human species. The ajetuink is fairly clearly responding to direct behavioural cues; it would be hard to make a case that its language abilities are any more advanced than, for instance, a sheepdog responding to whistles. The bufeget's counting could be explained as an example of the Clever Hans phenomenon, named after a horse in early twentieth-century Germany that could apparently work out sums. Investigation revealed it was responding to subtle non-verbal cues that guided it on when to stop counting. When these cues were blocked, the horse could not perform.

A way with words

The crinitex certainly seems to be operating on a different level from the other two species. It seems to be able start conversations without prompting; to use signs when humans are not present; to synthesise existing concepts into new ones, in a creative fashion; to use signs that can be interpreted by independent observers; to generate sentences with several clauses, displaying at least some degree of syntax; and to discuss abstract concepts way beyond food. Barring, perhaps, the crinitex's final comment, similar aspects of language use have been observed in chimpanzees brought up to speak using sign language.

Learning to fly

Leading linguist Noam Chomsky (b. 1928) is sceptical of such claims, asking: 'If apes have this fantastic [linguistic] capacity . . . then how come they haven't used it? It's as if humans can really fly, but won't know it until some trainer comes along to teach them.'

086 HIDING IN PLAIN SIGHT

Mallory had followed Principal Abacha around the entire school, and kept a close watch on him from assembly to the end of after-school clubs, and she had not once seen him eat anything – not even a biscuit! 'The principal's a vampire, or a ghoul, or maybe a zombie,' she said to Sanjay. 'I'm not sure which, but it must be one of them. It isn't natural, not eating anything ever. Sure, he sometimes drinks from a Thermos, but for all we know it's full of blood!' Sanjay began to object, but Mallory interrupted: 'We're going to spy on him at home, and catch him in the act.'

That evening they sneaked out of their bedrooms and met at the corner, then went two blocks up to a neat suburban house resembling all the others. The lights were on. They crept up to a side window and peered over the sill. Principal Abacha was sitting in front of his TV with a tray on his lap, eating a piece of chicken and some rice. Sanjay looked at Mallory. She looked confused, then horrified when a large arm reached out of the window and plucked her up by the collar. 'Excuse me, children, but what are you doing outside my window at eight o'clock in the evening?' Principal Abacha wanted to know.

When he heard their story, Principal Abacha asked them two questions: 'How many other teachers have you seen eating at school? Haven't you noticed me drinking my smoothies during the day?'

PIGEONS ARE A UBIQUITOUS FEATURE OF MODERN LIFE IN MUCH OF THE WORLD, ESPECIALLY IN TOWNS AND CITIES, BUT HOW IS IT THAT ONE NEVER SEES BABY PIGEONS?

Juveniles in disguise

Principal Abacha is making two points. Firstly, the children are unreasonably focused on his dietary habits without recognising that they rarely if ever see any of the teachers eating in public. Secondly, they actually *have* seen him eating, they just didn't realise it. Similar objections apply to the common trope that 'One never sees baby pigeons'. The claim is true of most garden birds; waterfowl are the only birds whose offspring are typically seen. Also, juvenile pigeons *are* common, but few people recognise them because they look very similar to adults. Wood pigeons without a white flash on the neck are juveniles, as are ordinary pigeons without shimmery greens and purples around the neck, but with dark eyes, long beaks and white bits at the top of the beak.

Private pigeons

Perhaps the main reason baby pigeons (squabs) seem notable by their absence is that, like Principal Abacha, pigeons prefer to keep some things private – specifically, child-rearing. Pigeons don't leave the nest until they have near-adult plumage, staying in the nest for up to 40 days – twice as long as most garden birds. The nests themselves are hard to see because pigeons, which descend from cliff-dwelling birds that evolved to nest on inaccessible rocky ledges, search for similar sites in the city. They will build their nests on covered ledges, under bridges or in roof voids, and they particularly like abandoned buildings.

If you did see a squab, you might wish you hadn't. They are notoriously ugly and resemble partially bald dodos, with outsized beaks too big for their heads.

087 A PROBLEM OF SCALE

Takashi is supposed to be project-managing an extension to his castle, but things are not going well. When the plans were revised to make the extension twice as long and twice as wide, Takashi ordered twice as many floor tiles as before. When they arrived, the foreman pointed out that he had only ordered half the required number; the length and width had doubled, but this meant that the area had quadrupled.

When the engineer said they needed to double the size of the central pillar, Takashi, burned from his previous experience, had ordered four times as many bricks. Once again, the foreman pointed out that he had under-ordered; since volume is the product of height × width × length, when the pillar's dimensions had doubled, its volume had increased eightfold. So when the foreman queries how much extra weight the enlarged pillar can support, Takashi confidently asserts that since it has eight times the volume of the previous design, it can bear eight times as much weight. To his dismay, the pillar crumbles and the castle falls down.

Big-boned

Takashi's doomed castle helps to illustrate why giant monsters such as King Kong or Godzilla are unlikely because of physical limits on animal size. Galileo (1564–1642), in his landmark book *Two New Sciences* (1638), discoursed on this very topic. He explained that a bigger animal obviously needs bigger bones to support its weight, but that carrying capacity does not increase in proportion to the volume of the bone, rather in proportion to its cross section. So in order to support four times as much weight, a bone must be eight times bigger. A greater and greater proportion of the mass of a giant animal must be taken up by the very bones it needs to support this mass, which themselves weigh more and more. The law of diminishing returns thus limits the size of animals, at least on land.

The giant dinosaur paradox

Fossil remains prove that giant reptiles did once walk the land. Titanosaurs are estimated to have been at least 37m (120 feet) in length, weighing 70 metric tons (77 tons). Bulk on this scale seems to controvert Galileo's square–cube rule limiting the size of animals. Possibly the creatures mainly kept to swamps where their bodies could be supported by water.

'Nature cannot produce . . . a giant ten times taller than an ordinary man unless by miracle or by greatly altering the proportions of his limbs and especially his bones, which would have to be considerably enlarged over the ordinary.'

– Galileo Galilei, *Two New Sciences* (1638)

088 AS THE BEE FLIES

In the year 2044, the International Olympic Committee has a problem. How will it cope with the latest enhanced or 'meta' human to emerge from the laboratory? New gene-editing techniques have produced a metahuman who combines human genes with those of a honeybee. Although the 'Beethlete' – as the media insist on calling her – cannot fly, she does possess the remarkable endurance of the bee, and is able to power her feats using the same, entirely legal natural foodstuff upon which the bee relies: honey.

The Beethlete may be the ultimate long-distance swimmer: Just as a bee is able to cover distances of up to 13.5km (8.4 miles), which is 900,000 times its own body length, so the Beethlete is able to swim up to 900,000 times her own body length. Since she is 1.72m (5 feet 8 inches) tall, this means she can cover a distance of 1,548km (962 miles). The energy in 1g (0.035 ounce) of honey can fuel roughly 1,400 bee-kilometers (870 miles), about 95 million times the insect's own body length, so it should take no more than 0.01g (0.00035 ounce) of honey to cover 13.5km (8.4 miles). The equivalent amount for the Beethlete, who is roughly 600,000 times bigger than a honeybee, would be about 6kg (13.2 pounds) of honey. Unfortunately this is bad news for the bees, because making this amount of honey requires bees to visit roughly 24 million flowers and cover 1.06 million km (660,000 miles).

'Hornets and wasps . . . are devoid of the extraordinary features which characterise bees; this we should expect, for they have nothing divine about them as the bees have.'

– Aristotle (c. 384–322 BCE)

As far as it needs to

According to beekeeper lore, a bee can fly as far as it needs to. The purest test of the bee's distance limit would be to put a hive in the middle of the desert with a ring of flowers at a set distance, then gradually increase this distance and see how far out the ring can be moved before the bees can no longer reach it. In fact versions of this experiment have been done, and they found that bees can forage for up to 11.26km (7 miles), but that if they have to travel more than 6.4km (4 miles) they burn more fuel than they can gather and the hive loses weight. Separate research has put the maximum flight distance of single honeybees at up to 13.5km (8.4 miles), while the tropical solitary Euglossine bee, *Euplasia surinamensis*, has been observed flying up to 23km (14.3 miles).

HONEYBEES ARE CELEBRATED FOR THEIR INDUSTRY, AND ARE OFTEN SAID TO TRIUMPH OVER THE LAWS OF PHYSICS SIMPLY BY TAKING WING. JUST HOW FAR CAN A BEE FLY?

089 SWIMMING ON MARS

The first Martian explorers had been amazed to find subterranean caverns filled with lakes of a liquid new to science, a form of the extremely dense metal osmium. Now tourists visiting the Red Planet are routinely briefed on the peculiar dangers of the osmium lakes, but no one ever pays much attention to the safety briefing.

Luiz was having a great time on Mars. He had climbed to the top of Olympus Mons and base-jumped off the side of the Valles Marineris canyon. Now he and his friends were exploring the great caverns with their spooky metal lakes. Their environment suits kept them perfectly protected from the extreme temperatures and unbreathable atmosphere. They were equipped with enough air for a two-hour excursion, and it only took ten minutes to get back to the rover on the surface, so Luiz figured they still had another half hour or so before they had to leave.

'Watch this!' he called to his friends, skimming a rock across the silvery lake. It hardly made a splash, and instead of sinking it lay half-submerged. 'Cool,' whistled Luiz. 'This stuff is weird. I wonder what it feels like?' He climbed down to the rim of the lake and touched the surface with his foot. 'Luiz, stop!' called one of his friends, but he paid no heed. 'Look,' he shouted, 'somebody could seriously float in this.'

FLIES CAN WALK UP WALLS AND EVEN ON CEILINGS, AND THERE ARE PLENTY OF INSECTS THAT SKATE AROUND ON WATER AS THOUGH IT WERE ICE, SO ONCE IT'S FALLEN INTO THE WATER, WHY CAN'T A FLY GET OUT OF THE TOILET?

'Wait, Luiz, they said not to—' But it was too late: Luiz had jumped in. The liquid was so dense, he floated with more than half his body out of the lake. The strange silvery tar oozed across his suit, coating and soaking the fabric. 'I'm swimming,' he laughed, splashing about. 'Very funny, now get out, we've got less than 15 minutes to get back to the rover,' said his friend, unimpressed. 'OK, OK, let me get over to the side.'

Luiz splashed awkwardly to the edge of the pool. The thick liquid made it hard to move, and his arms felt as though they had heavy weights attached. When he got to the side he tried to haul himself out, but it was as if someone were dragging him back down. The more he splashed, the more of the gloopy tar coated his suit and the harder it was to move. Luiz started to panic.

Heavy metal

Luiz is similar to a fly in a toilet: The simple maths of density, surface area and volume is against him. On Earth, a man getting out of a bath is covered with a film of water about 0.5mm (0.02 inch) thick, with a total weight of about 0.5kg (about a pound). Compared to his size and strength this is insignificant, but it becomes more significant in proportion to body size the smaller you get. A wet mouse will be carrying approximately its own weight in water, while the water covering a wet fly will significantly outweigh it. The liquid osmium in which Luiz is splashing is far denser than water, so that a coating of it weighs more than he does, making it impossible for him to hoist himself out.

'As everyone knows, a fly once wetted by water or any other liquid is in a very serious position indeed.'

– J. B. S. Haldane, *On Being the Right Size* (1929)

090 FIGHT CUBS

'Ladieeees and gentlemen! It's FIGHT NIGHT! Welcome to the most anticipated showdown in the animal kingdom, as we seek to settle once and for all the true champion of the big cats, the undisputed heavyweight King of the Animals!

'In the Blue Corner, fighting out of Sundarbans National Park in West Bengal, is the stripy-faced assassin, the black-and-orange butcher, the most deadly solitary hunter of them all . . . the Bengal tiger!

'And in the Red Corner, fighting out of the Kruger National Park in South Africa, is the maned monster, the roaring rampager, the biggest and baddest bully on the plains . . . the African black-maned lion!'

The bell rings, and they're off – Round One! The two big cats circle each other warily. The paw from the south, with his shaggy mane, is giving up nearly 15kg (33 pounds) to his larger rival. To begin with there's little sound, as each of these proud warriors attempts to stare down the other. And now they're vocalising! Hisses are giving way to deep growls, and then to tremendous roars from the lion.

Round Two, and the phony war is over. Both animals come out swinging, up on their hind legs, each attempting to gain a height advantage over his rival. The tiger strikes first, using his superior strength to knock down the lion, and right away he goes for the killer choke hold, clamping his jaws around the lion's neck. But what's this? The lion's shaggy mane means his rival simply cannot get a good hold, and now the tiger is shaken off. He retreats to his corner as the bell goes.

Now comes the third and final round. The lion, looking more confident now, feints with his left paw, putting the tiger off balance; the lion slashes at him, now he's on top of him, and oh my, that looks like a knockout blow. The lion has crushed the tiger's windpipe with a single clash of his mighty jaws and, folks, it looks like it's all over. It appears that the mane may have swung the odds in the lion's favour, and now we have a winner! The lion is the true king of the jungle!

Tale of the tape

Lions and tigers do not overlap anywhere in their range – even enclaves of Asiatic lions do not overlap with tiger territory. Possibly it was a different story in the past, and there have been many staged or accidental fights in history, from the Roman arena to modern menageries, zoos and circuses. Expert opinions differ, and the evidence from historical fights is not always clear, but on the whole the lion seems to outmatch the tiger, thanks to a number of advantages:

- Lions are social creatures, and males in particular are used to fighting rivals, so a wild lion is likely to have had more fight 'training' and experience than a tiger, which is a solitary creature.
- In the wild, tigers usually go to some length to avoid fights with rivals, whereas lions are generally more aggressive.
- The shaggy mane of the lion provides a natural defence against the main attack techniques of the tiger, which are the choke hold around the neck and the disabling attack on the spine.

091 DANCING ON THE CEILING

Jeff is a huge fan of Lionel Richie. As the ultimate tribute to the four-times Grammy Award winner, Jeff intends to actually dance on the ceiling. His first attempt involves some suction cups, but he has to be rescued when he is unable to move his feet or reach them to undo the cups. For his second attempt he researches the feet of the common housefly, paying particular attention to the tarsal claws, large hooks on the end of each foot. Attaching crampons to his own feet, Jeff finds that he is unable to gain any purchase on the minute cracks and irregularities in his ceiling, although he does succeed in making several holes.

Next, Jeff learns that the pads, or *pulvilli*, on the bottom of a fly's foot are covered with thousands of tiny hairs with spatulate ends, known as *setae*. Using a waffle iron and some quick-setting latex, Jeff attempts to cast his own setae-covered pulvilli. Although with care and hard work he is able to create a sole for his shoes that has hundreds of ridges and bumps, he is unable to pack in as many as the fly and he finds that his specially cast soles have no special adhesive properties. Once again he lands on his backside.

Seeking further clues, Jeff flash-freezes a piece of glass on which a fly has been walking and notices tiny, greasy footprints. Inspired, he applies a thin layer of treacle to his latex soles. Tests show that they are indeed remarkably sticky, able to adhere to the ceiling for long periods. Brimful of confidence, he straps on his shoes and, with the aid of some scaffolding, affixes his feet to the ceiling. For a short, glorious moment he sticks there, but then he falls off.

FLIES SEEM TO CRAWL ABOUT WITH IMPUNITY ON ANY SURFACE WHATEVER ITS ANGLE. HOW CAN A FLY WALK UPSIDE DOWN?

Four to the floor

Jeff has almost achieved his dream of clinging to the ceiling like a housefly. His final error was to try to stick on using his feet as the only points of adhesion. A housefly uses all six of its feet to gain adhesion, and when walking upside down ensures that at least four feet are in contact with the surface at all times.

Stuck on you

A fly also has the obvious advantage of weighing a fraction as much as Jeff, and thus needs far less adhesive power in order to stick. In fact, one of the fly's main problems is getting its feet unstuck. The tarsal claws can help, acting like crowbars to pry the foot free.

No glue needed

While flies and most (perhaps all) insects can secrete oily substances to make their feet sticky through capillary adhesion, geckoes eschew this trick altogether. As with the fly, the tips of the gecko's toes are covered in thousands of tiny folds and ridges (called *lamellae*), each of which is covered with tiny setae or hairs. These present a massive combined area of contact between the gecko and the climbing surface, across which operates the electrostatic attractive force known as Van der Waals attraction. Van der Waals is a weak force, but across such a large area it adds up, making the gecko's foot sticky enough to support many dozens of times its own weight.

092 CRUNCHY BITES

The theme of this year's school fete was international cuisine, and Thanh decides to showcase at his stand some distinctive street food from his ancestral homeland of Vietnam: grasshoppers. He boils the grasshoppers, dips them in egg and deep-fries them into a delicious crunchy snack.

Disappointingly, people turn their noses up at his offerings; they pull faces and say 'yuck', preferring the stands on either side, that are serving hamburgers and chicken, respectively. Some of them are surprised Thanh is even allowed to serve insects – aren't they dirty and unhygienic? A few people even think they must be poisonous.

Finally, Thanh decides he has had enough. He climbs onto a chair and makes a loud sales pitch. 'Who here wants to eat healthily?' he asks. Everyone puts a hand up. 'Who wants a low-fat nutritious food packed with more protein than any meat you can buy?' People murmur approvingly. 'Who here cares about the environment, wants to limit global warming? Who wants to conserve water, land and biodiversity?' People shout agreement. 'Who wants a solution to famine and food insecurity for billions of people?' There are claps and cheers. 'Then come to my stand and eat grasshoppers!'

Protein packed

Compared to other animal protein sources, insects generally have a much better balance of protein to fat and carbohydrate. Grasshoppers and crickets can contain as much as 20 percent protein (even more when dried) – almost as much as beef, but with only a quarter of the fat content and about 40 percent of the calories. Insects may also be rich in polyunsaturated fatty acids such as linoleic and ∝-linoleic essential fatty acids. They are low in cholesterol and high in calcium and iron.

Sustainable food

The UN's Food and Agriculture Organisation recommends insects as a potential solution to several problems around sustainability and food production. As the world's appetite for protein increases, as developing nations bring their dietary aspirations in line with the developed world, and the global population increases, livestock farming puts increasing pressure on already overstretched land use and environmental constraints. Livestock are also a major contributor to greenhouse emissions and water stress and, more fundamentally, they are inefficient at transforming feed into protein. Insects offer solutions to all these problems:

- There are about 40 metric tons (44 tons) of insects for every human on Earth.
- Grasshoppers, for instance, are around 20 times more efficient as a source of protein than cattle.
- Crickets need six times less feed than cattle to produce the same amount of protein.
- Crickets produce 80 times less methane than cattle for the same protein output.
- It is a myth that insects are unhygienic, especially when they are prepared correctly, and in fact the risk of pathogens may be lower than with livestock.

093 PLANET OF THE TREES

The men from the Global Forestry Commission surveyed the barren plain. Until the previous week there had been a thick forest here, but it had been levelled with astonishing speed. The first man got out a big map of the world and used a marker pen to black out the area they were surveying. Highlighted in green, the map showed global tree cover from the end of the last Ice Age. Half of the green area was now covered in black ink.

The lady from the logging company pointed out that her industry plants hundreds of millions of trees a year. But the men from the Forestry Commission weren't impressed. Trees are being harvested and habitats destroyed at such a colossal rate that, despite all the planting, the net annual loss of trees is still 10 billion. And we can't afford to lose them, the Forestry men said. Trees soak up carbon dioxide, shelter biodiversity, hold soils together and prevent flooding, not to mention generate oxygen.

The logging lady suggested that if the industry scaled up its replanting efforts, perhaps they could plant a new tree for every one that had been cut down since the last Ice Age. The Forestry men retorted that even if a new tree were planted every second, it would take 96,000 years to make up the difference.

ROUGH ESTIMATES OF TREE NUMBERS HAVE TRADITIONALLY BEEN MADE USING SATELLITE IMAGERY, BUT ONLY IN 2015 DID A MORE ACCURATE MODEL FINALLY PROVIDE AN AUTHORITATIVE ANSWER TO THE QUESTION, HOW MANY TREES ARE THERE?

Counting trees

In 2015 a team under Thomas Crowther of Yale University developed a new model for estimating global tree numbers, combining hundreds of detailed ground-sourced datasets regarding forest densities with satellite imagery to arrive at the most accurate figure ever obtained. They were astonished to find that there are around 3 trillion trees on the planet, eight times higher than the previous most authoritative estimate. They also calculated that this represents around half of the number present before the rise of human civilisation.

Outnumbered

Despite there being only half as many trees as there were 11,000 years ago – not to mention 7 billion more people since then – trees still outnumber humans by 420 to 1.

'For in the true nature of things, if we rightly consider, every green tree is far more glorious than if it were made of gold and silver.'

– **Martin Luther (1483–1546)**

094 ISLAND LIFE

'We'll be safe here,' Adam assured Steve as they stepped ashore onto the barren rock. Steve had developed an intense phobia of all types of animal, so they had come to settle on a brand-new island, a previously underwater volcano that had recently risen above sea level. After waiting for the island to cool, Adam and Steve had arrived on their yacht and set up camp, secure in the knowledge that they were the only living things on the island, barring microbes.

The next morning they were woken by a squawking noise, and looked out of their tent to see at least a dozen seabirds walking about. A strong wind blew in from the east, the direction of the nearest landmass. Several flies and a wasp were seen, while Steve was horrified to notice gossamer threads of silk floating past, many bearing tiny parasailing spiders. Within the month at least a dozen species of insect were buzzing about, mostly where the seabirds had taken to nesting, and tufts of coarse grass were shooting up.

After a storm had passed over the island, Adam went down to the shore and noticed mats of tangled vegetation being washed up on the beach. Climbing off one of the mats was a small tortoise, while another one carried a tiny lizard. Adam decided not to tell Steve about the reptiles; with only one of each, at least there was no danger of breeding. The next week he was surprised to discover the lizard proudly sitting atop a nest of eggs.

HOW DO NON-FLYING ANIMALS COME TO BE LIVING ON AN ISLAND?

New arrivals

Adam and Steve were living in a fool's paradise, if they thought they could escape nature. Any newly opened ecological niche is rapidly colonised, and new islands are no exception. The mid-Atlantic volcanic island Surtsey, discovered peeking above the waves near Iceland in November 1963, hosted lichen, moss and flying insects by the following year, while birds visited it within just a few days of its appearance. By 1965 several types of plant were growing there, and many invertebrates, from ticks to earthworms, arrived on the wind or carried by birds.

Reptiles make the best colonists

The first non-avian land vertebrates to arrive on a new island are likely to be reptiles, floating on rafts of vegetation; tortoises may survive a sea crossing even without a raft. The relatively slow metabolism of reptiles makes them better suited to surviving such crossings, while their ability to reproduce parthogenetically (with only one parent) means that even a single colonist can found a population.

Island life

Island life can have strange effects on those animals that do make it across. Large animals tend to shrink in size, as with pygmy hippos and the now extinct pygmy elephant; smaller ones sometimes grow, as with Komodo dragons or the now extinct moa; and birds may lose the power of flight, as with the flightless cormorant of the Galápagos and the extinct dodo of Mauritius.

095 BIGFOOT

Wendy is sick of other people. She thinks humans are lame, and she is determined that she and her family will get away from normal society, fending for themselves in the wilderness.

The first place they move to is Myakka River State Park in southwestern Florida, in the United States, but since the park is only 19km (12 miles) long, Wendy finds it hard to avoid constant contact with tourists and the locals. She relocates the family to the forested mountains of Humboldt County in northern California. Although far from any settlements, Wendy finds that they are often disturbed by the activities of logging companies. Forced by the relative scarcity of resources to forage over a wide area, she inevitably encounters lumberjacks and forest rangers. Eventually there are complaints that Wendy and her family have damaged logging company property and stolen food, and that they have littered and spoiled areas where they have camped.

With a warrant out for their arrest, Wendy and family decamp to the vast wilderness areas of British Columbia in Canada. But keeping away from townspeople is not the same as avoiding determined trackers, and they cannot avoid leaving a clear trail of evidence. The police recover discarded rubbish, bits of torn fabric, spoor and footprints, and it proves easy for them to find Wendy's campsite, and eventually to photograph and capture Wendy herself.

THE 1958 DISCOVERY OF AN ENORMOUS HUMANLIKE FOOTPRINT IN HUMBOLDT COUNTY, IN NORTHERN CALIFORNIA, SPARKED AN ENDURING CRAZE FOR BIGFOOT, THE NORTH AMERICAN APEMAN. COULD SUCH A CREATURE REALLY EXIST?

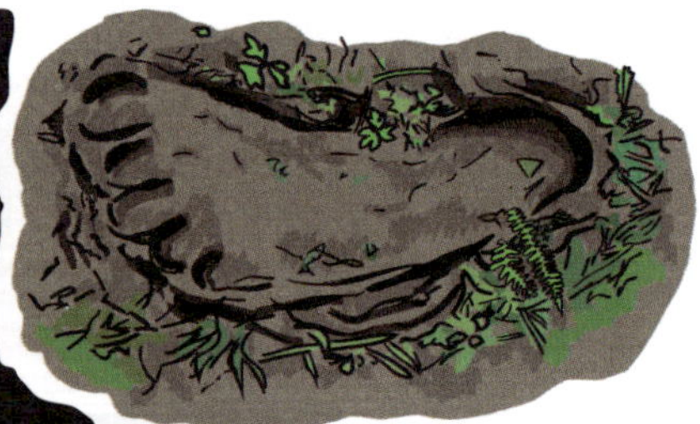

Hide and seek

Wendy has just discovered that even in the midst of a wilderness, it is hard to avoid accidental encounters with other people, and impossible to elude determined trackers. Needing to forage for food for herself and her family, Wendy had to venture out of her hiding places and range across a wide area, dramatically reducing her chances of remaining undetected. If Wendy wanted to establish her family line as an enduring population, she would have needed a considerable number of other people also living in the wild, and again this would significantly reduce the chances of remaining concealed. All of these same considerations apply equally to supposed populations of relic hominid, best known as Bigfoot, believed by some to be living in the wilderness areas of North America.

A Bigfoot for every state

All of the places that Wendy visits have been associated with Bigfoot sightings, including southwest Florida. In fact, Bigfoot-type sightings have been reported from nearly every state in the United States, as well as from dozens of countries in every corner of the globe. Bigfoot and his relatives must be extremely adaptable.

The case against

Even if sightings from more populous areas are discounted (and who is to say that one claimed sighting of a giant, unknown apeman is more credible than any other?), belief in Bigfoot still has many hurdles to overcome. For instance, in order for a hominid species to have survived to the present day, a sizable population must have been present at least up until around the 1950s. Yet no convincing hard evidence, in the form of hair or spoor samples, or bodily remains, has ever been found and verified by qualified, independent observers.

096 STICKY HEAT

It is the year 2100 and veteran geologist Moira Kuszpinski is working in the Persian Gulf state of Abu Dhabi. From her office she can see the deserted city streets outside. The blazing summer heat means that no one in the city ventures outside during the day except those who have absolutely no choice; everyone else stays in the air-conditioned interior spaces. But Moira knows that the luxury of air-conditioning is not available to hundreds of thousands of people in the surrounding countryside and desert, or to billions of people in regions of similar latitude and ocean proximity who are probably experiencing the same climate. The pity she feels for them turns to horror when she reads the weather forecast: The wet-bulb temperature predicted for next week is 35°C (95°F).

Moira is especially well equipped to understand exactly what this means, because she is one of the last people alive to have visited the Crystal Cave of Naica, in Mexico's Chihuahua Desert region. Long since flooded after the mine was abandoned and its pumps switched off, the Crystal Cave had been one of the wonders of the world, containing the largest crystal formations ever discovered. But only a handful of people had ever been inside it because of the conditions within; Moira remembered the heavy ice-pack suit and respirator she had had to wear simply in order to spend 20 minutes in the cave.

When she sees that the wet-bulb temperature forecast extends to most of the Southwest Asian tropics, Moira begins to tremble with fear.

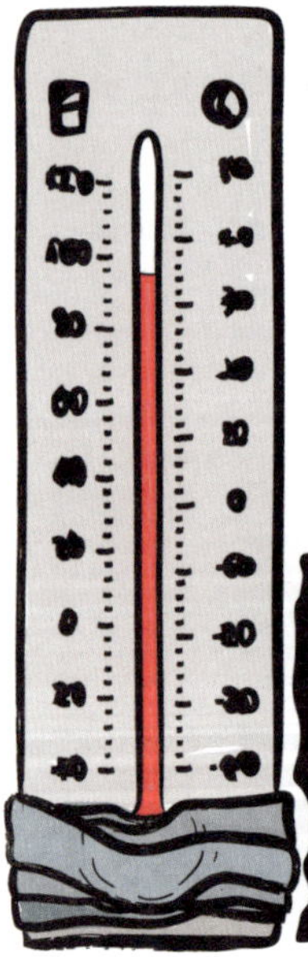

MUCH ATTENTION FOCUSES ON THE THREAT THAT GLOBAL WARMING MIGHT MELT POLAR ICE CAPS, BUT WHAT ABOUT NEARER THE EQUATOR? COULD GLOBAL WARMING MAKE PARTS OF THE PLANET UNINHABITABLE?

Wet-bulb temperatures

Normally, weather reports supply the temperature as recorded by a thermometer measuring the air, possibly alongside a separate measurement of humidity. What counts most for human health is the wet-bulb temperature. This is measured by a thermometer whose bulb is wrapped in a wet cloth, and therefore records the temperature of a moist surface that has been cooled as much as possible by evaporation.

When sweating fails

The normal temperature of human skin is 35°C (95°F). So long as the air surrounding the skin – and crucially, the air in contact with the moist surface of the lungs – has a wet-bulb temperature below this, water can evaporate from the body into the air. This is how the body cools itself, and this is why, given an adequate water supply, humans can withstand intense desert heat and even run marathons in it. However, if the wet-bulb temperature exceeds 35°C, moisture cannot evaporate from the lungs taking heat with it, and instead hot, moist air accumulates in the lungs, causing breathing problems and overheating.

The Crystal Caves

With temperatures of 55°C (131°F) and 100 percent humidity, the Crystal Cave of Naica is currently the only place in the world where the wet-bulb temperature exceeds 35°C. But according to several studies, including one published by Jeremy S. Pal and Elfatih Eltahir in *Nature* in 2015, global warming is projected to produce wet-bulb temperatures this high in parts of Southwest Asia by 2100. If this happens, anyone without access to air-conditioning will die in less than an hour.

097 NOBODY'S FRIEND

God is sitting in judgment on a number of glitches in Creation. The world was created in a hurry, and mistakes were made, but now He is working through a long snagging list. 'What's next on the docket?' He asks. The Archangel Raphael leads in a slightly sheepish-looking insect: 'Mosquito, Your Awesomeness. She's charged with spreading disease and misery across the globe.' God looks down His nose at the mosquito: 'Are you the same guys who swarm in those thick clouds on the tundra in the summer? Didn't I hear that a cloud of you actually suffocated a caribou?'

'In our defence, Your Immanence, we provide all manner of ecosystem services,' pleads the insect. Her defence counsel, the Archangel Gabriel, concurs: 'Mosquito larvae and adults are an essential element of many ecosystems. They provide a major dietary contribution to many species of bird and fish – there's even a species called the mosquito fish – and the aquatic larvae help break down detritus in water, keeping it clean.'

'That's true, Your Ineffableness,' said Raphael, 'although the seraphim tell me that other filter feeders would quickly fill the gap, and that predator species would soon adapt to feed on organisms that will flourish in the mosquito's absence.' The mosquito piped up desperately, 'But, Your Empyreal Highness, we do a lot of pollination too. I mean, without us the cacao plant wouldn't get pollinated, and there'd be no chocolate.' 'Er, actually,' corrected God, 'I believe it's the ceratopogonid midges that pollinate cacao.' The mosquito looked guiltily at the floor.

WOULD MOSQUITOES BE MISSED IF THEY VANISHED OVERNIGHT?

'So what you're saying,' summed up God, 'is that while they might cause wobbles in a few ecological niches, they're not exactly irreplaceable.' 'Yes, Your Sublimity, and the truth is no one would miss them very much if they were gone. Caribou hate them, and humans are always complaining: malaria, yellow fever, dengue fever, Japanese encephalitis, Rift Valley fever, Chikungunya virus and West Nile virus – all spread by mosquitoes.'

'My, my,' chuckled God, wagging His finger at the mosquito, 'you have been naughty. But don't look so worried. If there's one thing I've learned about my creation, it's that it tends to fill ecological niches pretty darn quickly. If I eradicated you from the face of the Earth, it wouldn't take long for some other vector to take over spreading all those ghastly diseases. Besides, I'm tired of extinguishing species: The humans do enough of that. Just promise to lay off the poor caribou.'

They'll none of 'em be missed

A surprising number of mosquito biologists and ecologists canvassed by the journal *Nature* in 2010 seemed to think that mosquitoes would not be particularly missed if they were to vanish overnight. There was acknowledgment that adults and larvae contribute to the diets of many species, and that larvae help break down debris in the water, while adults perform some pollination roles. But most experts believed that the void left if mosquitoes were to vanish would soon be filled by other species. According to entomologist Daniel Strickman of the US Department of Agriculture, 'The ecological effect of eliminating harmful mosquitoes is that you have more people. That's the consequence.' But nature despises a vacuum, and it is likely, just as God from our story suggests, that the disappearance of one vector animal would simply lead to the emergence of another.

098 KEEP IT IN THE FAMILY

Abeni, Bijan, Cheng and Danesh are playing a game. Abeni and Cheng play to win. They use tricks and stratagems. They watch carefully for any vulnerability and seize the chance to score wherever and however possible. At first, either Abeni or Cheng wins every round.

Bijan and Danesh are brothers. In the fourth round, Bijan has a chance to make a move that will take points off Danesh, but he doesn't take it. Later he makes a move that leaves him vulnerable to attack by Cheng, so that he will come last in that round, but which ensures that his brother will win the round.

In the fifth round, Danesh makes moves that benefit Bijan but not himself. Bijan wins the round. Abeni and Cheng protest; the brothers are helping each other and this is cheating. Danesh points out that the rules say nothing about whether players can help each other through legal moves. Abeni argues that such behaviour makes no sense; Danesh might win a round, but Bijan will come last. Where is the benefit to Bijan?

Bijan points out that, by making altruistic moves, he and his brother can ensure that one of them wins every round. So whatever happens, their family is winning. Cheng counters that the game offers no prizes for group or collective wins; only individuals can be declared winners. Danesh suggests that he and his brother will take it in turns to help the other, so that each wins alternate rounds, while Abeni and Cheng are shut out.

Abeni wants to know how Danesh can trust his brother to reciprocate his altruism. Bijan could accept Danesh's help in one round, and win, and then in the next round he could refuse to reciprocate, betray his brother and win that round too. Danesh explains that while Bijan could do this, since they are siblings his brother has an interest in not betraying him, especially if he wants the tit-for-tat of altruistic reciprocation to continue.

We are family

Abeni and Cheng represent the classical Darwinian thinking about how organisms play the game of evolution by natural selection. Individuals play for their own sake; anything that benefits others necessarily disbenefits them, and hence anything they do to help others, especially at their own expense, is a losing strategy. It was this perception of the workings of evolution that led to the dismissal of early research into plant communication, in which Jack Schultz and Ian Baldwin reported in 1983 that maple saplings can transmit, apparently via airborne pheromones, warnings about herbivore attack that can trigger receiving saplings to boost their defence systems proactively. Sceptics contended that expending energy to help out competitors makes no evolutionary sense. There is now evidence, however, that plants preferentially communicate with relatives, suggesting that Bijan and Danesh's strategy, known to evolutionary biologist as kin selection, is at work.

Fungal internet

In addition to signalling via airborne chemical messengers, there is also strong evidence that trees communicate via an extensive network of fungal threads (the mycorrhizal mycelia) passing between root systems, sometimes described as a sort of fungal internet. Trees signal each other across this network, and can even pass nutrients from one plant to another, so that older trees can help saplings and parents can help offspring trees.

MANY GARDENERS LIKE TO TALK TO THEIR PLANTS, BUT CAN THE PLANTS TALK TO EACH OTHER? CAN PLANTS COMMUNICATE?

099 TREAD CAREFULLY

'Alarm! Alarm! The Archons are attacking!' The cry rang out around the battlements of Castle Grimhold, and the brave Men of the South girded themselves for the onslaught. The Archons were nightmarish foes, with massive armoured bodies that skittered across the ground on dozens of spindly, claw-footed legs.

Baron Trueheart rallied his men: 'Never fear! Our foe cannot overcome the defences of the Great Outer Wall, which have stood fast against all invaders for a thousand generations.' The Archon army reached the foot of the Great Outer Wall and paused for a moment. 'See how they quail before its impenetrable protections. Our alchemists have coated it with Holdfast Tar; any of the monsters that attempt to scale it will be stuck fast like ants on honey.' Yet, even as he spoke, the Archons began to spew upon one another torrents of hideous grease. Once they were slick with the vile stuff, they began to clamber up the wall.

'Soap, bring us soap,' called Baron Trueheart. 'Empty the laundry houses and the baths, call out every goodwife and potboy in the citadel – we must have soap.' Quantities of soapy liquid were brought forth, and the Baron had them tipped down the wall, washing away the grease from the Archons' feet wherever they hit their mark. Many of the beasts became hopelessly fouled in the tar, but most carried on regardless. 'Fall back,' ordered the Baron, his forces retreating to the Inner Keepfast.

A QUARTER OF KNOWN SPIDER SPECIES ARE ORB-WEAVERS, WHICH CREATE THE CLASSIC SPIDER'S WEB OF POPULAR IMAGINATION, AND COAT PARTS OF IT WITH BLOBS OF STICKY GLUE. WHY DON'T THESE SPIDERS STICK TO THEIR WEBS?

'Be of good cheer, lads,' he reassured his troops. 'The walls of the Keepfast are more impenetrable still. A dense network of razor-encrusted wires has been slung across its face. None shall scale these walls!' Once more, however, the Archons paid little heed to his boasts, picking their way carefully but quickly up the Keepfast wall. 'See how they step between the wires,' called out one of the guards. 'Look,' cried another, 'they are using the pincers on the tips of their feet to keep the wires clear of their vulnerable parts.' Trueheart unsheathed his mighty blade. 'This is no time for admiring their anatomy, men, now we must FIGHHTTT!'

Washing spiders' legs

As Baron Trueheart is discovering, a well-engineered foot can overcome stickiness. The Archons have evidently picked up a few tricks from spiders, which use a variety of methods to avoid becoming stuck to their own webs. One of these was discovered back in 1905, when French naturalist Jean-Henri Fabre (1823–1915) theorised that orb-weaver spiders, which coat their capture threads with blobs of glue, avoided adhering to their own webs by coating their limbs in a sort of oil. To test his theory Fabre washed the spiders' legs with solvent and claimed that this made them much more likely to get stuck. In 2011, researchers at the Natural History Museum in Bern, Switzerland, repeated Fabre's experiment under controlled conditions, and found that solvent-cleaned spiders legs will indeed stick to their own webs.

Fancy footwork

Spiders' feet are covered in tiny hairs, minimising the area of contact between spider and web, and they have *tarsi*, or barbs, on their feet, which grip the web so that it is held away from the rest of the foot.

100 ANIMAL OLYMPICS

The Animal Olympics produces some surprising winners. The feline race is won not by the cheetah, but by the domestic cat. The humble household pet then goes through to the land animals final, where it finds itself lapped four times by the tiger beetle.

In the race for sky-diving vertebrates, pre-tournament favourite the peregrine falcon is beaten into a poor third by both the swallow, in second place, and the Anna's hummingbird, which is the clear winner. In fact, the cameramen filming the tournament have a hard time keeping pace with the Anna's hummingbird, despite the fact that one crew is on board a fighter jet with its afterburners on, while another crew has been stationed on the Space Shuttle during a re-entry burn. Both are left for dead by the hummingbird.

However, in the Grand Final even the Anna's hummingbird is left trailing by the Animalympian Champion of Champions, the copepod. The copepod is a tiny shrimplike creature that floats in the ocean. Probably the most abundant multicellular animal on the planet, the copepod also proves to be both the fastest and the strongest.

The last event in the Animal Olympics is a race in which the human gets to race the copepod. The result is humiliation for the human: By the time he has completed one lap, the copepod has completed over 296!

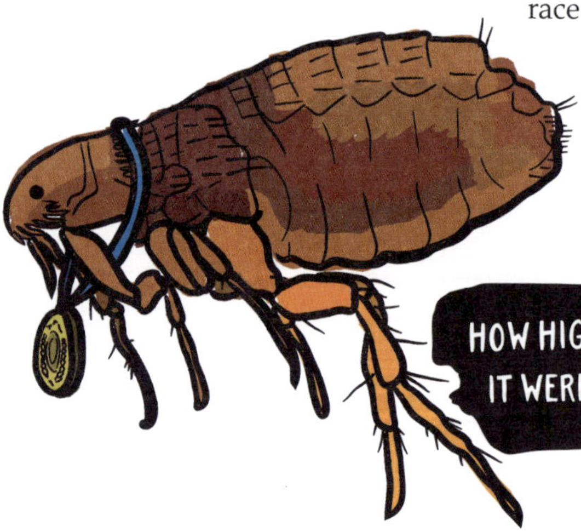

HOW HIGH COULD A FLEA JUMP IF IT WERE AS BIG AS A HUMAN?

Handicapping the Animal Olympics

These results were obtained by handicapping the contestants in the fairest way possible. Obviously the animals in question vary enormously in size, by several orders of magnitude. How then, can the speed of a cheetah be compared with that of a copepod? To show their speed *relative to their body sizes*, each animal is ranked according to how many multiples of its own body length it can cover per second. For instance, the top speed a human sprinter can achieve is about 11m/s (36 feet per second). Assuming a height of around 1.8m (5 feet 11 inches), this gives a relative speed of ~6 body lengths per second (blps). The copepod is about 1mm (0.039 inch) long, but can move at speeds of up to 6.4km/h (4mph), giving it a speed of roughly 1,780 blps.

Unexpected champions

Matching up animals by blps produces some unexpected outcomes. The cheetah is the fastest land animal, achieving speeds of up to 113km/h (70mph), but in terms of blps the cheetah (25 blps) comes in behind its domestic cousin the house cat (29 blps). Faster than either of these is the tiger beetle, which can run at 1.9km/h (1.2mph), or approximately 125 blps. The fastest known vertebrate – a stooping peregrine falcon – moves at 200 blps but a swallow can dive at 350 blps and the Anna's hummingbird dives at speeds of up to 385 blps, making it faster than both a fighter jet with its afterburners on (150 blps) and the Space Shuttle during re-entry (207 blps).

High jump

A cat flea can leap up to 20cm (8 inches), despite being less than 3mm (an eighth of an inch) long, meaning it can jump about 160 times its own body length. If the flea were blown up to human size, assuming it was not required to cope with the square–cube size law (see p. 191), it would be able to cover up to 290m (950 feet) in a single bound.

101 SOAKED TO THE BONE

Spongegirl's filthy foes have combined to create a particularly fiendish trap, by turning her own cleansing superpowers against her! The perfidious plotters have somehow duplicated her own body tissues, and by foul craft constructed long tunnels from them. In order to reach their lair and put a stop to their wickedness, Spongegirl must now pass through the tunnels.

The first of them is spongy, much like her own body, but almost as dry as a bone. As she squeezes through the constricted space, Spongegirl's soft, squishy torso is crushed against the dry sponge, and her precious bodily fluids leak out, soaking into the dry matter surrounding her. The only way she can maintain her own fluid levels and avoid drying out herself is to swallow as much of the surrounding sponge as possible. Her internal processes are able to gather what little moisture is contained within the dry sponge, but embarrassingly she also has to excrete the large amounts of sponge she has had to gobble.

Finally she makes it to the next tunnel, also made of sponge, but this time sopping wet. As she squeezes through the soaking tunnel, her own spongy body soaks up water at a tremendous rate. Although she doesn't need to eat any of the surrounding sponge, she once again finds herself embarrassed as she is forced to urinate copiously. By the time she reaches the villains' lair, she is in a very bad mood. 'It's time to mop up some dirt!' she warns them.

LEAVING ASIDE ANY SPECULATION ABOUT THE SUBJECTIVE EXPERIENCES OF A FISH, IT IS NEVERTHELESS POSSIBLE TO ASK, IN PHYSIOLOGICAL TERMS, 'DO FISH GET THIRSTY?'

Osmosis in action

Spongegirl is analogous to a fish in different environments. The first tunnel is comparable to seawater. Though not actually dry, seawater does have a higher concentration of salts than the body fluids of a bony fish. Osmotic pressure thus means that water is drawn out of the fish's body, dehydrating it, just as Spongegirl loses water to the dry sponge. Just as she has to eat the sponge to reclaim the lost water, so a saltwater fish has to drink seawater, and in this sense it can be said to get thirsty. Where Spongegirl excretes sponge, a saltwater fish has to excrete salt.

Rivers of wee

A fish in freshwater is analogous to Spongegirl in the second tunnel. The fish's bodily fluids have much higher salt concentrations than the surrounding water, and so water migrates into the fish's body, following the osmotic gradient. Freshwater fish do not need to drink, and so presumably do not get thirsty, but like Spongegirl they do need to urinate a lot.

Teetotal sharks

Despite living mostly in seawater, sharks and other elasmobranch fish do not get thirsty. They retain organic molecules that most other vertebrates excrete, such as urea and trimethylamine oxide, and this increases the concentration of their bodily fluids, so there is no osmotic gradient with seawater and they do not lose water.

FURTHER READING

Adam, John A. *A Mathematical Nature Walk*. Princeton, NJ: Princeton University Press, 2009.

Alter, Torin, and Robert J. Howell, Eds. *Consciousness and the Mind–Body Problem: A Reader*. Oxford, UK: Oxford University Press, 2011.

Austad, S. N. 'Menopause: An Evolutionary Perspective.' *Experimental Gerontology* 29 (3–4) (May–August 1994): 255–63.

Bonner, John Tyler. *Why Size Matters: From Bacteria to Blue Whales*. Princeton, NJ: Princeton University Press, 2006.

Bostrom, Nick. 'Are You Living in a Computer Simulation?' *Philosophical Quarterly* 53 no. 211 (2003): 243–55.

Brunning, Andy. *Why Does Asparagus Make Your Pee Smell?* Berkeley, CA: Ulysses Press, 2016.

Bryson, Bill. *A Short History of Nearly Everything*. London, UK: Doubleday, 2003.

Cappellini, G., Y. P. Ivanenko, R. E. Poppele, F. Lacquaniti. 'Motor Patterns in Human Walking and Running.' *Journal of Neurophysiology* 95 no. 6 (June 2006), 34–37.

Chown, Marcus. *Quantum Theory Cannot Hurt You: Understanding the Mind-Blowing Building Blocks of the Universe*. London, UK: Faber & Faber, 2014.

Comins, Neil F. *What If the Earth Had Two Moons?: And Nine Other Thought-Provoking Speculations on the Solar System*. New York, NY: St Martin's Press, 2010.

Corsini, Raymond J., and Alan J. Auerbach, Eds. *Concise Encyclopedia of Psychology* (2nd edn.). Chichester, UK: John Wiley, 1996.

Crowther, T. W., et al. 'Mapping Tree Density at a Global Scale.' *Nature* 525 (2015), 201–5.

Enders, Giulia (trans. David Shaw). *Gut: The Inside Story of Our Body's Most Under-Rated Organ*. Melbourne, Australia and London, UK: Scribe, 2016.

Erikson, Erik H. *Identity, Youth and Crisis*. New York, NY: W. W. Norton, 1968.

Fang, Janet. 'Ecology: A World Without Mosquitoes.' *Nature* 466 (2010), 432–4, 2010.

Fowler, Michael. *Galileo and Einstein Home Page*. 2006. galileoandeinstein.physics.virginia.edu

Futuyma, Douglas. *Evolution* (3rd edn.). Sunderland, MA: Sinauer Associates, 2013.

Gleick, James. *Time Travel*. London, UK: 4th Estate, 2016.

Goldacre, Ben. *Bad Science*. London, UK: Harper Perennial, 2009.

Gordon, Jeffrey I., et al. *Extending Our View of Self: The Human Gut Microbiome Initiative* (HGMI), 2005. www.genome.gov/pages/research/sequencing/seqproposals/hgmiseq.pdf

Gregory, Richard L. (ed.). *Oxford Companion to the Mind* (2nd edn.). Oxford, UK: Oxford University Press, 2004.

Gribbin, John. *In Search of Schrödinger's Cat*. London, UK: Black Swan, 2012.

Gross, Richard D. *Psychology: The Science of Mind and Behaviour*. London, UK: Hodder Education, 2015.

Hogg, Paul. 'The Mpemba Effect: Does Hot Water Freeze More Quickly Than Cold Water?' *The Mole*, Royal Society of Chemistry, 1 (January 2012): 5.

Holmes, Richard. *The Age of Wonder: How the Romantic Generation Discovered the Beauty and Terror of Science*. HarperPress, 2009.

Jha, Alok. *The Water Book*. London, UK: Headline, 2016.

Joachim, David, and Andrew Schloss. *The Science of Good Food*. Toronto, Canada: Robert Rose, 2008.

Levy, Joel. *Really Useful: The Origins of Everyday Things*. Richmond Hill, Ontario, Canada: Firefly Books, 2002.

– – . *History's Greatest Discoveries And the People Who Made Them*. New York, NY: Metro Books, 2015.

– – . *Why We Do the Things We Do*. London, UK: Michael O'Mara, 2015.

– – . *The Infinite Tortoise*. London: Michael O'Mara, 2016.

Libbrecht, Kenneth G. *Snow Crystals*. 2016; snowcrystals.com

Locke, John. *An Essay Concerning Human Understanding*. Oxford, UK, and New York, NY: Oxford World's Classics, 2008.

Lundin, Knut E. A.'Non-Celiac Gluten Sensitivity – Why Worry?' *BMC Medicine* 12 (2014): 86.

McGee, Harold. *On Food & Cooking: An Encyclopedia of Kitchen Science, History and Culture*. London: Hodder & Stoughton, 2004.

Mlodinow, Leonard. *The Drunkard's Walk: How Randomness Rules Our Lives*. London, UK: Penguin, 2009.

Nave, C. R. *HyperPhysics*. 2016. hyperphysics.phy-astr.gsu.edu/hbase/index.html

Niemitz, Carsten.'The Evolution of the Upright Posture and Gait: A Review and a New Synthesis.' *Naturwissenschaften* 97 (3) (March 2010): 241–63.

O'Neil, Dennis. *Human Biological Adaptability: An Introduction to Human Responses to Common Environmental Stresses*. 2013. anthro.palomar.edu/adapt/Default.htm

Onstott, Tullis C. *Deep Life: The Hunt for the Hidden Biology of Earth, Mars & Beyond*. Princeton, NJ: Princeton University Press, 2016.

Pal, Jeremy S., and Eltahir A. B. Eltahir. 'Future Temperature in Southwest Asia Projected to Exceed a Threshold for Human Adaptability', *Nature Climate Change*. October 26, 2015. www.nature.com/articles/nclimate2833.epdf

Paulos, John Allen. *Innumeracy: Mathematical Illiteracy & its Consequences*. New York, NY: Hill & Wang, 1988.

Philbrick, Nathaniel. *In the Heart of the Sea*. London: HarperCollins, 2000.

Raymer, Dorian M., and Douglas E. Smith. 'Spontaneous Knotting of an Agitated String.' *PNAS* 104 no. 42 (October 16, 2007) 16432–7.

Roach, Mary. *Packing for Mars: The Curious Science of Life in Space*. London, UK: Oneworld, 2011.

– – . *Gulp: Travels Around the Gut*. London, UK: Oneworld, 2016.

Sheldrake, Rupert. *Dogs That Know When Their Owners Are Coming Home and Other Unexplained Powers of Animals*. London, UK: Hutchinson, 1999.

Slater, Alan, and J. Gavin Bremner, Eds. *An Introduction to Developmental Psychology*. Chichester, UK: John Wiley, 2011.

Sterling, Ian. *Polar Bears*. London, UK: A & C Black, 2012.

Stromberg, Joseph.'Why Do Mosquitoes Bite Some People More Than Others?' smithsonian.com, 12 July, 2013; www.smithsonianmag.com/science-nature/why-do-mosquitoes-bite-some-people-more-than-others-10255934

Summers, Adam.'Shoe Fly', *Natural History Magazine*, February 2006. www.naturalhistorymag.com/biomechanics/172099/shoe-fly

The Surtsey Research Society; www.surtsey.is/index_eng.htm (various articles).

Taylor, David. *The Life and Death of Stars*. 2012. faculty.wcas.northwestern.edu/~infocom/The%20Website/index.html

Tsuzuki-Hayakawa K., Y. Tochihara, and T Ohnaka.'Thermoregulation During Heat Exposure of Young Children Compared to Their Mothers.' *European Journal of Applied Physiology and Occupational Physiology* 72 (1–2) (1995): 12–17.

Weinstein, Lawrence, and John A. Adam. *Guesstimation: Solving the World's Problems on the Back of a Cocktail Napkin*. Princeton, NJ: Princeton University Press, 2008.

Wells, J'aime.'Language-Using Apes', *Philosophy Now* 89 (March/April 2012): 31–4.

INDEX